THE
MACKENZIE PIPELINE:
ARCTIC GAS
AND
CANADIAN ENERGY
POLICY

THE
MACKENZIE PIPELINE:
ARCTIC GAS
AND
CANADIAN ENERGY
POLICY

Edited by
Peter H. Pearse

McCLELLAND AND STEWART LIMITED

The Canadian Publishers
McClelland and Stewart Limited
25 Hollinger Road, Toronto

PRINTED AND BOUND IN CANADA

Carleton
Contemporaries

A series of books designed to stimulate
informed discussion of current and
controversial issues in Canada, and to
improve the two-way flow of ideas
between people and government.

ISSUED UNDER THE EDITORIAL
SUPERVISION OF THE INSTITUTE OF
CANADIAN STUDIES, CARLETON
UNIVERSITY, OTTAWA.
DIRECTOR OF THE INSTITUTE
A. D. DUNTON

TABLE OF CONTENTS

PREFACE

We in Canada are now on the threshold of some major discussions in the
energy field. We must soon decide at what rate we are to develop our frontier sources of oil and gas, with all the implications that such development
has for the lives of those that make their homes in those regions, for the
environment of those areas and for the national economy.

These are the words of Canada's Minister of Energy, Mines and
Resources in introducing the government's recent study of Canada's energy position.[1] This federal report, like those recently
undertaken by the provinces of Ontario, Quebec, Alberta and
British Columbia, arose out of a general uneasiness about future
national and provincial energy policies in Canada. Energy has
suddenly become a much more urgent issue. Our trade in oil and
gas is becoming a central concern in our relations with the
United States; the environmental impact of exploring for,
extracting, processing, transporting and burning fuels is a growing source of public apprehension; the foreign ownership of our
oil and gas industry is a prime target of those concerned for our
economic independence; the different interests of producers and
consumers are coming to a head in relations between provinces,

and between them and the federal government; and the energy sector is assuming a growing importance in our whole economic structure. Canadians are understandably concerned about whether the policies and institutions we have evolved for managing our energy resources are adequate to these new conditions.

The world energy picture is changing fast. Other industrial nations – particularly the United States – are caught up in a so-called energy crisis. The expansion of supplies from traditional sources has not kept pace with the inexorable growth in demand. The great oil producing states of the Middle East are making ever-increasing demands of foreign producing companies and importing countries, raising new concerns about the security of supplies from abroad. Enormous efforts are being made to develop technologies for new energy sources. But in the forseeable future the industrial countries must rely mostly on conventional energy fuels from home or abroad, and the cost is rising.

In this scene, Canada stands alone among western industrial nations in having energy resources apparently surplus to forseeable domestic needs. We can take advantage of the rising world demand for our natural resources by expanding exports, which would not only produce new flows of foreign exchange but also generate new economic activity through expansion of production facilities. Or, we can save our energy reserves for our own future needs, and by price and export controls we can insulate Canadians from the rising costs attending the energy crisis of other nations. The arguments for both courses of action are subjects of vigorous public debate, and, as the federal government admits, the national interest in these issues has yet to be determined.

As in so many Canadian policy issues, the position of the United States is a major consideration. It is already clear that the United States government regards Canada as a valuable source of secure energy supplies.

> The United States should work diligently with Canada to reach a continental energy policy that assures our mutual security. Such a policy should cover energy broadly and should deal with not only oil but natural gas, coal and hydroelectric and nuclear sources.
> Pending agreement on such a policy, which may take several years to negotiate, Canada and the United States should develop an effective mechanism to permit an orderly growth of imports of oil and natural gas from Canada.[2]

Our close neighbour and largest trading partner, with by far the biggest markets in the world, thus appears ready to buy huge quantities of products that we have plenty of.

How should Canada respond? The traditional reaction of Canadian governments, and of Canadians generally, to new export opportunities has been one of unequivocal approval. But here the issue is not lumber, or wheat, or automotive machinery – but energy. And the prospect of exporting vast amounts of energy today conjures up deep apprehensions among Canadians – even among those who, in other circumstances, advocate a policy of freer trade. Whether it is in the interests of Canada to export substantial quantities of energy depends upon the balance of benefits and costs that Canadians would incur. Both the gains and the sacrifices are exceedingly complicated. The benefits include not only the revenues that the exports would generate but also a variety of secondary stimuli to the Canadian economy. Because of the magnitude of these impacts, and the fact that they will be spread through many future years, the effects are difficult to measure and are subject to a good deal of uncertainty. The costs include not only the direct requirements of enormous amounts of capital and labour for development and transportation, but also the less easily measurable (but nonetheless important) adverse implications for the natural environment and economic dislocations to other sectors of the economy. Whether other effects can be counted as benefits or costs is uncertain: the impact of frontier activity on native communities, of foreign capital inflows; of price changes that effect producers and consumers (and regions) differently – these depend on the form of development and the importance we attach to them.

Thus the subject of energy resource development impinges on many of the most sensitive issues in modern Canadian society. Will exporting energy aggravate or ameliorate these problems? The debate is just beginning.

All these issues are coming to a head in proposals to export recently-found reserves of natural gas in the Mackenzie Delta, near the arctic coast in the Northwest Territories. If current plans are realized, Canadian authorities will soon be presented with a proposal to construct an enormous pipeline up the Mackenzie Valley to carry gas. from the Delta to markets in the United States and eastern Canada. The multi-billion dollar project is being planned by a consortium of twenty-six companies with interests in oil and gas reserves, transportation and marketing.

This project is related to the development of large United States oil reserves on the north slope of Alaska – at Prudhoe Bay, some three hundred miles northwest of the Mackenzie Delta. There, the oil developers propose to construct a hot crude oil pipeline southward across Alaska to the ice-free port of Valdez, and there to ship the oil by huge tankers to refineries on the United States west coast. The whole development has been delayed by legal actions sponsored by environmental groups in the United States and Canada, and there has been considerable pressure (including some apparently ambiguous overtures to Washington from the Canadian government) for an oil pipeline across Canada to the United States mid-west markets instead. However, (at the time that this was written, at least) the United States Congress appears to have now cleared the way for the trans-Alaska pipeline-tanker route for Prudhoe Bay oil. But there is natural gas mixed with the oil in the underground reservoirs at Prudhoe Bay, and this is where a Mackenzie Valley gas pipeline enters the picture. The only alternative to the expensive process of liquifying Alaskan gas for tanker shipment is to deliver it by pipeline, overland across Canada. A pipeline of some 2500 miles can be economically viable only if it can deliver very large quantities of gas. Thus the proposal will involve a huge pipeline, four feet in diameter, capable of delivering more than four billion cubic feet of compressed gas each day to southern markets. To fill this enormous capacity, it is proposed that half the gas will be drawn from Canadian reserves in the Mackenzie Delta, and half from the oilfields on the arctic coast of Alaska. The pipeline alone would cost in excess of five billion dollars – a project larger than anything Canada has yet undertaken. It's impact on Canada's north would be staggering.

The Mackenzie Valley gas pipeline proposal has provoked a sheaf of engineering and environmental studies, economic analyses and public policy reviews on a scale that is claimed to surpass anything in our history. However, it is questionable whether this volume of studies will leave the Canadian public uniquely well informed, because for the most part their purpose is not public education. Rather, they are sponsored by investors, environmentalists and tax collectors attempting to overcome our ignorance of the complex technical, ecological, sociological and economic consequences peculiar to Northern resource development. As a result, they are fragmentary both in their coverage and in their point of view.

Moreover, it is doubtful that these studies will enable Canadians to comprehend the range of choices open to them, and how these choices can be made. The frustration felt by many Canadians about the apparent inevitability of northern resource development, and their anxiety about whether it will take place to our national advantage, is expressed by Richard Rohmer:

> ... we just don't have the centrality, the force, the direction, or the policy-making overview that is necessary to make or give effect to a national goal. Nowhere is this more apparent than in the inability of Canada to come to grips on an overview basis with the uncontrolled but controllable monster I have chosen to call the Arctic Imperative, which is taking off like a wild rocket going in all directions concurrently. The people who are in control of the Arctic Imperative, if anyone is, are the American oil and natural gas firms to whom we have sold the commodity in the ground for a pittance, and the gigantic American natural gas and oil distribution firms which control the market.[3]

One does not have to agree that we have abdicated control of northern development to American corporations to conclude that Canadians are ill-prepared for the extremely important decisions they will have to make during the next couple of years in response to proposals from the oil and gas industry.

This volume is designed as a contribution to the debate on northern energy development, concentrating on the Canadian public policy aspects of this question. The authors prepared their papers during the summer of 1973 with the information available at that time. Circumstances are changing rapidly, of course. Within a few days of writing this (September 1973) the federal government has proposed major changes in Canada's National Oil Policy; the Supreme Court of the North West Territories has recognized aboriginal rights which strengthen Indian land claims over vast areas of the region of particular concern to us here; the oil-rich Arab states have threatened new strictures on supplies to the United States; and the United States government has announced its intention to lift its price ceiling on natural gas. And almost every month brings new information about the quantity and kind of energy reserves in the north and elsewhere, the form that the proposed pipeline project would take, its environmental, economic and political implications, and the policy alternatives available to us. The contributors to this volume were forced to resist the temptation to continuously revise their work in response to new changes; otherwise this book could never have appeared.

But the reader must be cautioned that new developments will continue to alter the basis of their analyses and hence also their conclusions.

The authors reflect a variety of points of view about northern energy development. While each paper (except Chapter 10) has been exposed to comment and criticism from the other contributors, we have considered it more desirable to present contrasting viewpoints rather than a uniform position. Thus, we hope, the reader can gain not only an appreciation of the issues at stake, but also an understanding of the different sides to the public debate.

Clearly, the question of developing arctic gas through a Mackenzie Valley pipeline must be viewed in the light of Canada's general energy situation – the position of energy in our economy and in our foreign trade, our domestic needs, the framework of policy and institutions that regulate the industry, and so on. The introductory essay by Anthony Scott and Peter Pearse attempts to provide this perspective, and thus to set the stage for examination of the arctic gas issue. The second paper, by Earle Gray describes, and presents the case in favour of, the Consortium's proposal for a Mackenzie Valley gas pipeline – the focus of attention in the rest of the book. This is followed with an analysis, by Paul Bradley, of the major "players" in Canada's energy scene, and how the varying interests of producing companies, governments and consumers interplay in directing energy development and in sharing the benefits. Ernst Berndt then provides a brief examination of the problems in assessing Canada's future energy needs.

The next four essays deal with more specific issues raised by the pipeline proposal, each written by an authority in the subject. Andrew Thompson and Michael Crommelin examine the unique arrangements governing exploration, tenures, royalty payments and leasing in the Canadian north and their implications for Crown revenues from exploitation. Stuart Jamieson's paper addresses the problem of the impact of development on northern Indian and Eskimo communities. The environmental impact of resource development and pipeline construction is assessed by Everett Peterson. And John Helliwell describes the results of his rigorous analysis of the pressures that pipeline construction and operation would bear on Canada's exchange rate, trade balance, employment and other conditions. Chapter 9, by Milton Moore, offers a general review of the options facing Canada. The final paper, Chapter 10, was prepared after all the others were in hand. It is the result of a decision on the part of some of the contributors to try to bring all the available quantitative information about the proposed development together, to provide as clear an indication as possible of the economic and other gains that Canadians could expect from the proposed development, and from alternative courses of action.

This collection of essays originated with a group of researchers at the University of British Columbia, some of whom attended a meeting of Canadian and United States petroleum exports in Vancouver in October 1972 at which the mutual interests of the two countries in oil and gas trade were discussed. That meeting revealed an unexpected divergence in the Canadian and United States points of view. It also left the Canadian academic participants convinced that the development of arctic natural gas was about to become a critical issue for Canadian national policy; that well-articulated analysis of the public policy questions at stake were not readily accessible to the Canadian public; and that some joint contribution to these problems would be useful. We therefore undertook to prepare a series of papers on the important aspects of energy policy in Canada, focusing on the implications of a Mackenzie valley pipeline to exploit northern natural gas. To supplement and balance the resources available at the University of British Columbia, we invited additional contributions from Mr. Gray, a spokesman for the Consortium that is currently preparing the pipeline proposal and from Dr. Peterson, a government expert in environmental effects of arctic development.

We circulated rough drafts of the papers for discussion at a series of evening meetings organized by the University's Centre for Continuing Education in downtown Vancouver in early 1973. At these meetings, each author had the benefit of critical comment from other contributors as well as from a lively group of interested outsiders. The revised papers that appear here owe much to those who participated in the discussions, and we are grateful to them, as well to those who helped to organize the meetings. Preparation of the manuscripts for publication was made possible by a financial contribution from the University's "Intellectual Prospecting Fund", and to the anonymous donor of that fund we want to express our gratitude.

PHP
VANCOUVER
September, 1973

FOOTNOTES

1. Honourable D. S. MacDonald, *An Energy Policy for Canada: Phase I*, Department of Energy, Mines, and Resources, Ottawa, 1973, p. v.
2. *The Oil Import Question: A Report on the Relationship of Oil Imports to the National Security* ("Schultz Report"), by the Combined Task Force on Oil Import Control, U.S. Government Printing Office, Washington, D.C., 1970, p. 362.
3. Richard Rohmer, *The Arctic Imperative*, McClelland and Stewart, Toronto, 1973, pp. 217-18.

1.
THE POLITICAL ECONOMY
OF ENERGY DEVELOPMENT
IN CANADA

Anthony Scott and Peter H. Pearse

Canadians must soon decide where their national interest lies in
the development of arctic energy resources. Proposals are now
being prepared for construction of an enormous pipeline up the
Mackenzie Valley and across the western prairies to link reserves
of natural gas on the arctic coast with markets in the south–
primarily in the United States. It would be a huge project; larger
than anything yet attempted in Canada. And it promises to
bring to a head all the emerging questions about Canada's
energy policy.

Yet many of the issues raised by the proposed Mackenzie
Valley gas pipeline are similar to those of other great development projects we have experienced in this century. The Columbia
River Treaty, the Trans-Canada Pipeline, the St. Lawrence
power and seaway project are obvious precedents; and there
have been others involving water resources, coal and minerals.
These large natural resource development projects have much in
common. Their huge scale invokes unusual requirements for
capital, and the spending of these enormous sums not only
produces profound changes in the economy of the region of

development but also causes repercussions throughout our domestic and external trade. Their size means also that a major transformation of the natural environment takes place. In addition, as a result of these projects' significant economic and political implications, governments have typically been deeply involved. Often a foreign government – usually that of the United States – has stood behind the project's developer or facilitated its access to product and capital markets by regulating prices, imports, security issues or the location of works. Typically, also, more than one Canadian government has been involved, with a province's proprietory interest at least as influential as the federal power over trade and foreign relations. Finally, nearly all these great projects have been concerned with development and sale of energy.

In many respects, then, the Mackenzie Valley project raises issues that the Canadian economy and political system have experienced before. What makes this proposal more interesting – more crucial – in our opinion, is the global energy situation in which the project is proposed. Canadians must make up their collective mind about this project at a time when the United States and all the other industrial nations of the Western World are alleged to be in the grip of an "energy crisis."

A. CANADA AND THE ENERGY CRISIS

Economists are hesitant to use the term "energy crisis" because it implies a breakdown in essential energy supplies. The situation is more accurately described as one in which the growth in energy consumption is more rapid than the rate at which new supplies are forthcoming from traditional sources *at today's prices*. This growing scarcity presents a crisis insofar as it will have far-reaching effects on the whole world by creating a new pattern of international interdependence. It will affect the trade among nations; relations between the great powers and their relations with poorer nations; almost every government's balance of revenues and subsidies; the future of the world's largest private corporations; and almost everyone's use of energy-consuming equipment, products and services. It also has profound (and to a large degree unknown) implications for the natural environment – the ecological balance and stability of the oceans, the atmosphere and northern tundra – and for the environmental and social well-being of frontier communities and crowded cities.

A "crisis" means literally, a turning point; an interval of suspense in the groping for new directions during which there is a danger that the new scheme of things will bring quite different economic and political relationships among nations, producers, consumers and social groups. Such crises are rare in today's world because the complexity of modern industrial economies, with their robust equilibrating markets and trade flows, alternatives for production and substitutability of products, makes them resilient to shocks and pressures, just as an ecosystem is more stable the more complex its structure. Great social crises normally result only from political revolutions such as those that have transformed Eastern Europe and China or from economic events of the magnitude of wars that penetrate every part of the economic fabric.

It is doubtful that the emerging scarcity of some sources of energy is of such a catastrophic nature. It is true that the pessimistic authors of *Limits to Growth,* and other apocalyptic works, raise the possibility that energy supplies might ultimately preclude continued economic expansion, but even they suggest that continued growth would more probably be limited by food shortages or environmental barriers. Nor do we need to take issue with the generations of scientists who have warned that increasing industrialization and population growth are hastening the exhaustion of our planetary energy stocks. All authorities are agreed that the present energy crisis is not to be interpreted as the beginning of the end of the world. Indeed, the forecasts of huge import requirements to meet American energy demands in the next few years also suggest that the development of new sources will enable that country to become self-sufficient in energy again in a couple of decades.

Instead, the emerging energy crisis is essentially a problem that is internal to the economies of the United States and the other advanced Western nations. Traditional sources of energy cannot continue to expand as they have in the recent past to meet growing demands. Adjustments must be made. And if new arrangements are not made carefully and frugally they can have adverse consequences for both producers and consumers.

The crisis, and the means of resolving it, cannot be understood through economic analysis alone. It requires also an appreciation of the positions of various political interest groups and how they stand to be affected. Energy-policy decisions will produce different distributions of population and wealth, and new centres of political power. The hegemony of oil-producing regions like Texas in the United States and Alberta in Canada is likely to be challenged by the rise of new producing areas, changes in export and import policy, and the development of new sources of energy. All these changes will be propelled not only by technological advances and market forces but also by the interplay of producers' organizations and consumer groups striving to secure their own future by seeking favourable decisions about prices, imports, taxes, subsidies and so on. Thus the energy crisis is a study in "political economy" – a term from Classical and Marxian economics which describes how economic forces influence social class and income structures in capitalist societies and invoke challenges and responses in political power. The rest of the world cannot escape the changes that will take place in the industrial nations now facing the energy crisis. The patterns of international trade, the relative value of currencies and the prices of imports and exports will all be affected, and other countries will feel these impacts to a degree that depends largely upon their involvement in world trade.

Foremost among these is Canada, the only industrialized nation apart from the Soviet Union which can be considered to have reserves of energy surplus to her domestic needs. Traditionally, Canada has been an importer of coal and oil, and an exporter of coal and hydropower, but recently she has become a net exporter of almost all forms of energy. As a net exporter, the energy crisis is less ominous for Canada than if she were a net importer, dependent on the crumbs from the rich man's table. Indeed, other nations' crises can be regarded by Canadians as an auspicious trend, creating new opportunities for serving a strong export demand. Or, if we should choose, we have the option to detach ourselves from the American energy problem and conserve our reserves instead. Detachment can be expensive, however. It involves not only sacrificing potentially large private incomes and public revenues, but also the risk that surplus resources conserved for the future will lose their export value as United States technology develops and Americans adapt to new sources of power and heat.

To understand Canada's new policy choices, we must first attempt to understand the elements of the energy problem in the United States. This will help to explain the "excess demand" for Canadian natural gas. We then turn to the political economy of energy in Canada to examine why United States demand – not-

withstanding enormous reserves in Canada's western provinces –
reaches far north to the Mackenzie Delta and the arctic.

B. ENERGY SUPPLY AND DEMAND IN THE UNITED STATES

Ever since the United States has been a dominant world power,
Washington has put heavy emphasis on the strategic value of
self-sufficiency in energy supplies. Until recently, this objective
could be met without serious difficulty. With enormous reserves
of coal, a steady rate of discovery of oil and gas, many opportunities for hydropower development, and an emerging nuclear-
electric generating capability, American imports of energy were
confined to certain regions and were more-or-less marginal to
total requirements. But in the last few years this situation has

TABLE I
United States Energy Supply 1970-1980 (heat equivalent)
(Trillion Btu)

Domestic	1970	%	1980	%
Oil	21,048	31.0	24,323	23.7
Gas	22,388	33.0	18,600	18.1
Coal	13,062	19.3	19,928	19.4
Hydro	2,677	3.9	3,033	3.0
Nuclear	240	0.4	9,490	9.3
Geothermal	7	–	343	0.3
Sub-total domestic	59,422	87.6	75,717	73.8
Imports				
Oil	7,455	11.0	22,984	22.4
Gas	950	1.4	3,880	3.8
Sub-total imports	8,405	12.4	26,864	26.2
Total Demand	67,827	100.0	102,581	100.0

Source: National Petroleum Council, *U.S. Energy Outlook*, Vol. II,
cited in *Energy in Ontario: The outlook and Policy Implications*, Vol. II, Advisory Committee on Energy (Toronto,
1973), p. 20.

changed. Demand continues to grow but supplies of oil and gas from traditional sources, which now serve three-quarters of United States energy consumption, are beginning to decline. Authoritative forecasts indicate that by 1980 the United States will be forced to import half of its oil requirements and some 18 per cent of its natural gas needs.[1] Table 1 summarizes United States production and imports of various forms of energy in 1970, with forecasts for 1980. Note in particular the expected expansion in imports of oil and gas during this decade. Later, import requirements can be expected to decline as new "unconventional" sources of power and heat are developed.

The current energy crisis in the United States is not simply a result of growing demand. Although consumption continues to grow at more than 4 per cent per year, this growth has been steady, and largely foreseen. This crisis arises mainly from unforeseen changes in the *supply* of energy sources.

The "shortage" of natural gas, which has recently attracted a good deal of attention, is in large degree a result of official ceiling prices. With the price held at a low level, sales have been stimulated and new supplies discouraged.[2] Consequently, existing reserves have been depleted and new reserves have not been developed. Indeed, a significant proportion of the natural gas withdrawn in the course of oil production is still burnt off (flared) at the wellhead because it is unprofitable to deliver it to markets. Another result of the low price ceiling is that production has been diverted from the regulated consumer market to the unregulated industrial market where higher prices have prevailed. The artificially low price explains much of the decline in production and the "shortage" reported when potential new customers have been refused supplies.

Government and industry agree that relaxation of these price controls would stimulate new exploration and development of gas resources in the United States. Accordingly, President Nixon's *Energy Message* of April 1973 contemplates allowing the price of newly-developed gas reserves to rise as high as market forces dictate, while maintaining the regulated price for existing reserves. These prices can be expected to diverge significantly, and utility companies will pay different prices for gas depending upon its source. Final consumers will pay a price that reflects the mix of supplies purchased from regulated and unregulated sources by utility companies. In short, gas prices will rise; average prices slowly, and the price for new stocks very sharply. This will have two predictable effects: (i) with higher gas prices, American consumers will substitute other fuels, especially oil, and (ii) there will be a very strong incentive for the gas industry, and industrial consumers, to contract abroad for foreign reserves – especially from Canada.

The switch to oil and other fuels cannot be rapid, however. The reserves in established oil producing regions of the United States appear to be finally declining, in spite of strong incentives to promote new exploration and development afforded by the United States government in the form of tax privileges and protection from imports. Crude oil prices in the United States have been maintained well above the cost of imported supplies from Venezuela and the Middle East. Yet efforts to discover new domestic supplies have yielded successively fewer, smaller and more expensive reserves. Hence the United States oil industry is turning from traditional areas to the "frontier" regions offshore and in Alaska and to shale oil. But by 1985, these new sources will be able to satisfy only 10 per cent of United States demand.[3] In short, the United States must turn to imports, if oil is to continue to carry the heaviest burden of energy demands. As Table I indicates, the projected requirements for imported oil amount to nearly half of United States total consumption by 1980 – some 10.7 million barrels per day. This has enormous political and economic implications:

(1) United States government energy policy will become less concerned with domestic tax incentives and more absorbed in foreign relations with producing countries. In particular, it will be concerned with establishing secure sources of supply. Security relates to the political stability of source countries, but it also calls for a diversity of sources and a transportation network that is safe from interruption.

(2) The concentration of economic power in the United States petroleum industry will shift from the present mosaic of independents and domestic companies to the "International Majors," with overseas reserves and connections.

(3) The centres of activity and income generation will shift from inland producing regions to the seaboard and to overseas countries.

(4) Because of the growing concern for environmental protection, those foreign supplies which have the lowest content of sulphur and particulates will be favoured over

others, and recent frustrations experienced in finding environmentally acceptable locations for refineries are likely to become more acute.

The United States oil producers and refiners have closer connections with Canada than with overseas suppliers, and Canada is regarded as the most secure foreign source of supply for both geographical and political reasons. Hence the United States industry is anxious to obtain increased imports of Canadian crude. But most Canadian crude is not considered a low-sulphur product. And, in any event, Canada's oil reserves in excess of domestic requirements are not expected to be sufficient to meet more than a small fraction of United States import requirements. These considerations sharpen the American demand for Canadian natural gas; it is available in large quantities, is the cleanest of fuels, and is secure and already integrated with the U.S. industry.

The energy crisis centres on the growing scarcity of the major fuels – oil and gas – in certain traditional supplying areas; but there are substitutes for both of these in many uses. In particular, they can be replaced by electricity, generated by falling water or by boilers fueled by coal or nuclear energy. But these other sources carry their own problems, and these problems aggravate the energy crisis. Opportunities for hydropower have to a large extent been fully exploited already, and this source cannot be expected to make a significant contribution to expanded energy needs. Nuclear generation holds much more promise: this source may provide 60 per cent of electrical energy and 30 per cent of total energy by the end of the century.[4] But the rate of adaptation to nuclear power is constrained by problems of safety from explosion, leakage and disposal of radioactive and thermal wastes. These problems have created enormous difficulties in finding acceptable sites for nuclear facilities. And so, while expansion of nuclear generation seems inevitable in the long run, both the problems associated with its production and the limitations of electricity as a substitute for other fuels in transportation and heating will prevent this source from supplanting a high proportion of oil and gas needs in the short run.

Vast quantities of coal are available at low cost, and coal has a convenient and well-understood technology. But political resistance to expansion of coal use is now greater than for other fuels. In the first place, the extraction of domestic coal – chiefly by open-pit methods – leads to increasingly unpopular environmental havoc in producing regions. Second, the burning of most types of coal produces more air pollution than does other fuels. And finally, rail transportation of coal is an expensive, dirty and disruptive business per energy unit compared to the transportation of oil and gas.

In summary, the United States energy crisis is the result of declining reserves of petroleum in the established domestic producing regions, artificially depressed prices for natural gas, and a growing environmental concern that obstructs expansion in thermal electricity supplies – all occurring in the context of a determination to become self-sufficient in energy. The United States must look increasingly to foreign supplies of gas and oil for the next 25 years, until self-sufficiency can be restored.[5] In securing the enormous American requirements for imports from abroad, the posture of the cartel of oil producing states (OPEC), the stability of the Middle Eastern countries which contain vast oil reserves, the Arab-Israeli conflict and relations with the Soviet Union will all play their part. While imported oil will probably be less costly than that from new domestic sources, it will also be much less secure. And its insecurity will lead to new efforts in domestic exploration and production and recourse to other fuels.

C. NEW PROBLEMS FOR CANADA

In this context, increased supplies from Canada appear especially attractive to United States governments and American suppliers, and so the full blast of avid American demand turns onto Canada. For example, some United States interests see Canada taking the place of the United States as a coal exporter in world markets, and some metallurgical industries see Canadian locations as particularly advantageous for heavy-energy processes such as smelting. More directly, American suppliers, now searching abroad for secure but low-cost sources give increasing attention to Canadian hydro, crude oil, and especially to Canadian gas – secure, clean, and relatively inexpensive.

Faced with these potential demands, Canada's problems – Canada's energy crisis – are quite different from those of the United States or Europe.[6] How should the reserves that are surplus to domestic needs be managed? They can be exploited to take advantage of the new export demand, but exporting is a matter not only of amount but also of timing. We can withhold energy from export not only to satisfy today's Canadian consumers, but also to make the best possible provision for both future Canadian consumers and for even more profitable future export opportunities. Or Canadian consumption of domestic supplies can be expanded to reduce the present dependence on imports, or to insulate Canadian consumers from rising costs elsewhere. Thus Canada's decision about the optimum quantity of energy to keep in reserve is impelled by a variety of overlapping motives, all relating to our national welfare.

How are these various considerations to be combined and weighted in making our export decisions? In this paper, and in others in this volume, much attention is centred on the "economic rent" – or net value generated in excess of all costs – that can be expected to result from a certain policy. An expected stream of rents over a number of years can be expressed in terms of a lump sum "present value" by discounting future amounts at an appropriate interest rate, thus enabling us to consider timing carefully and to allow for the diminished importance that people attach to postponed events. By calculating how these rents would be shared among producers, consumers and governments we can estimate the distribution of the gains among sectors, as is done in Chapter 10 of this volume.

Thus an estimation of the sectoral rents provides the essential economic information about the gains to be expected from development or export decisions. But there are other issues to be considered also. Environmental protection is becoming increasingly important: strip-mining, valley flooding and oil spills are as damaging here as they are in the United States. And there are other considerations – federal-provincial relations, regional development, the problems of northern native peoples, the balance of international trade and foreign ownership – all of which would be affected by expanded exploitation and export of energy resources.

In many respects, past energy policies appear inappropriate to the new demands on Canada's resources. Provincial and federal governments that have hitherto granted tax incentives to gas, oil and coal, subsidies to coal production and export, generous land alienation arrangements to petroleum exploration and development, and careful sharing of the limited export demand among the competing resource locations (which were often characterized by excess production capacity), must suddenly reverse direction. Now export limitations seem more appropriate than incentives, and higher royalties politically more attractive than subsidies.

But Canada cannot simply put on the brakes and withdraw from these new demands. The traditional exporting regions are still locally dependent on foreign sales and many see their future in further expansion. British Columbia and Alberta, once exporters of coal and hydro-electricity, are once again stressing coal, as well as oil and gas. Quebec seeks to become a large hydro exporter, and the coal of Nova Scotia and the Atlantic provinces may soon give way to large reserves of off-shore oil and gas. Thus, for governments, there is little political support for insulation from the American market. This is particularly true of the Canadian north. The economic development that coal, oil and gas have brought to eastern and western provinces now seems to be within the grasp of the residents of the northern territories. There, newly-discovered reserves of natural gas offer an independent source of natural resource income capable of eliminating their political and financial dependence on Ottawa.

The problem that Canadians face, then, is how to adjust their political response and internal policies in the face of these changed circumstances, in order to make the best of new opportunities for profit from exports while limiting social and environmental damage and assuring adequate supplies for domestic needs. This is the context within which the Mackenzie Valley proposal must be evaluated.

D. ENERGY AND THE PETROLEUM INDUSTRY IN CANADA

Canadians consume more energy per capita than any other nation except the United States. North Americans, with 6.5 per cent of the world's population, produce one-third of the world's total economic output and consume more than 37 per cent of the world's total energy production.[7] Energy consumption in Canada has recently been growing faster than in the United States, and faster also than the rate of growth of the Canadian Gross National Product. At 4 to 5 per cent per year, Canada's rate of growth in energy consumption has been among the highest in the western world.

While Canada consumes about 3 per cent of world energy production, this country has until recently consumed more than it produced. In 1965 Canada exported the equivalent of 25.6 million metric tons of coal in various forms of energy (almost all to the United States) and imported 56.5 million metric tons equivalent (over half from the Caribbean, a little less than a third from the United States and most of the rest from the Middle East). The reversal in Canada's foreign trade balance occurred rapidly between 1966 and 1971, when exports rose much more rapidly than imports.

The form of energy consumption in Canada, indicated in Figure 1[8], differs from most other western countries. Solid fuel (mainly coal) is considerably more important in most other industrial countries than it is in Canada. Over half our consumption is in liquid fuel, which is relatively high. The share occupied by hydro-electricity though small, is the highest among major countries, and the proportion of natural gas is higher than in any other country except the United States. Within Canada's energy sector, the production of oil and natural gas have grown particularly rapidly in recent year.

Figure 1 illustrates the growth in Canadian consumption of various sources of energy, with projections to the year 2000. It should be noted that total energy consumption is expected to continue to grow rapidly, and that oil and natural gas are expected to account for an increasing share of the total.

Natural gas production began as an offshoot of oil extraction in Alberta where gas was found associated with oil in underground reservoirs. With increased discoveries, and the construction of pipelines, gas production became an important sector of the petroleum industry serving distant markets with high-quality fuel. Today, Canadians consume a little more than a trillion cubic feet (t.c.f.) of natural gas per year, and another trillion was exported in 1972. Proven remaining reserves in the Western provinces exceed 52 t.c.f., of which some 14 t.c.f. are already under contract for export. More is expected to be "proven up" as development proceeds, and these figures do not include the substantial discoveries in the Mackenzie Delta and Arctic islands which are not yet connected to markets.

The petroleum industry, in Canada and elsewhere in the western world, is characterized by certain organizational and institutional features that have important implications for public policy. In the first place, most of the world's petroleum is produced by a few very large corporations. These huge companies epitomize the modern multinational firm, each with operations in most countries outside the Communist block where resources are found or where large markets exist. They can, and do, move their opera-

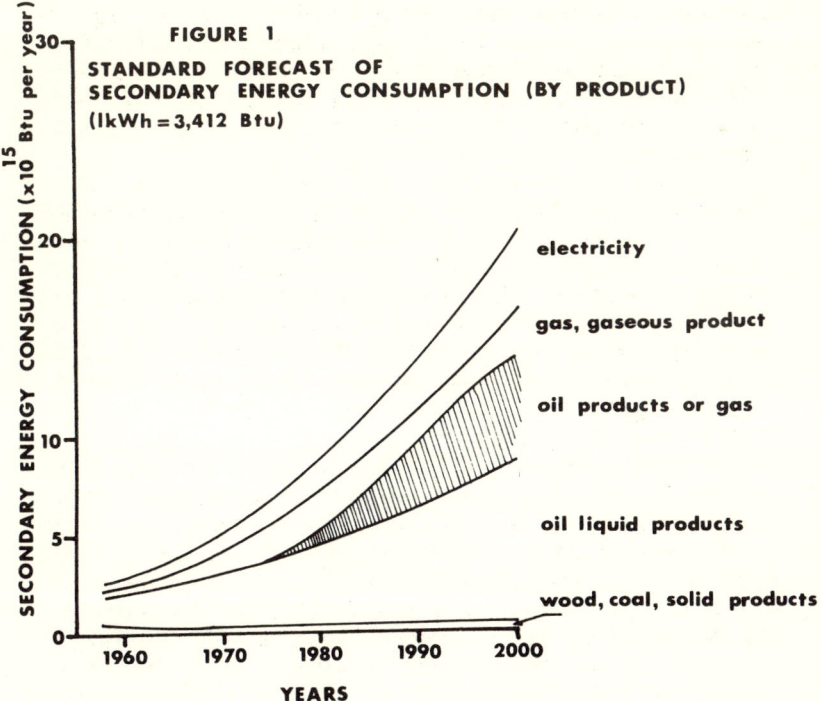

FIGURE 1

STANDARD FORECAST OF SECONDARY ENERGY CONSUMPTION (BY PRODUCT)

(1kWh = 3,412 Btu)

SECONDARY ENERGY CONSUMPTION ($\times 10^{15}$ Btu per year)

electricity

gas, gaseous product

oil products or gas

oil liquid products

wood, coal, solid products

YEARS

Source: Department of Energy, Mines and Resources, *An Energy Policy for Canada: Phase 1*, Vol. I, Information Canada, Ottawa, 1973. p. 73.

tions freely among nations, and are highly responsive to each economic and political environment in which they operate. This is one reason why a government must always design its policies affecting the industry in the light of the policies of other nations.

This handful of corporations includes some of the largest corporations in the world. Of the seven "International Majors," which control nearly three quarters of the refining capacity in the non-Communist world, five are owned and controlled in the United States.[9] With operations in more than 90 countries, they accounted, in 1966, for about 40 per cent of all United States overseas investment. Sales of the five United States Majors in 1967 was equal to a third of Canada's Gross National Product, and they are among the largest 12 industrial corporations in the United States. Nearly all crude oil and gas production, and much of the oil and gas marketing in Canada is controlled by these International Majors. Foreigners are reported to own 99 per cent of the petroleum industry.[10] In the current debate over the extraordinarily high degree of foreign ownership and control in Canada, this industry is usually cited as among those in which Canadians have least proprietary interest and managerial control.

A second source of public concern arises from the structure of the industry. Exploration, production, refining and marketing is in the hands of a small number of very large horizontally and vertically integrated firms which have all the characteristics of an oligopoly – homogeneous products, close cooperation in many activities, and the absence of vigorous price competition. The inflexibility of product prices and pervasive over-capacity in production and retailing are, partly at least, explained by this industrial organization.

The petroleum industry is conspicuously capital intensive. Most of the enormous quantities of capital it requires is imported, and these large international flows have an important impact on Canada's capital markets and foreign exchanges. The ratio of labour to capital used is lower than in almost any other industry, so that investment, per dollar, does not create much direct employment.[11] As Chapter 8 of this volume shows, a project of the magnitude of the Mackenzie Valley pipeline requires foreign capital inflows at rates that can have a marked impact on Canada's balance of payments and foreign exchange value, although the direct labour employment they generate is rather modest.

Finally, a most direct form of public involvement in the industry results from the fact that the underground resources, in western and northern Canada, are almost entirely under Crown title, and become private property only upon extraction. Thus provincial and federal governments have a vested interest in the pattern and efficiency of exploration, development and marketing, and in the royalties, rent and other charges paid to the Crown. Before turning to the particular issue of development of Mackenzie Delta gas, it is useful to note how all these factors interplay in determining governments' policies affecting energy in Canada.

E. ENERGY POLICY

Throughout history, energy has seldom, if ever, been recognized as a unique and unified problem of public policy. A good deal of attention has been paid, of course, to particular fuels and uses of energy, but energy, *per se,* has been far less important to policy makers than has its component parts. Government intervention in energy matters has everywhere been fragmentary, dealing with particular industries or resources or markets in particular times and places, and often influencing energy development as a means to accomplish diverse public objectives. Public policy on energy is necessarily as inchoate as "transport policy," the "war on poverty" or "urban policy." The piecemeal measures that provincial and federal governments have taken are sometimes called "energy policy," but it is difficult to perceive or even conceive of a single cohesive approach that deals with all energy decisions in a consistent framework.

Thus although Canada (like other western countries) has had no articulate energy policy as such, it administers an enormous variety of national, provincial and local measures that affect energy production and consumption, in pursuit of diverse goals. Various measures have been taken to improve the performance of the market – for example, through regulation of pipeline construction and operation, railroads, shipping and public utilities – and to enhance competition through anti-trust legislation and regulation of selling practices. The performance of the market has been further aided by public participation in geological surveys and dissemination of information and regulations to prevent over capacity and wasteful rates of extraction. And governments have attempted to provide for the public interest where the market fails, in matters of national security, safety, and protection of the environment.

Coming at the energy decisions from the distributional angle, governments have intervened in a variety of ways to generate public revenues from resources, in systems of royalties, rentals, bonus bids for leases, and special modifications of income, capital-gains, tariffs and other tax systems. An even greater array of measures have been taken to redistribute these revenues among regions, governments, and economic groups. Regional redistribution is accomplished through arrangements such as those involving inclusion of resource royalties and rents in provincial income in calculating equalization payments. Certain kinds of producers benefit from tariffs, quotas and subventions which influence production and consumption of particular fuels or in particular regions—such as the Canadian coal policy and the National Oil Policy. A host of special income tax arrangements have similar effects. And consumers are directly affected by tax-free fuel to some users, special treatment of rural markets and some industries, and the rate-fixing policy of power utilities.

Equally lengthy lists of public interventions could be made under the headings of measures to stabilize markets, measures to conserve resources for the future, measures to promote growth and regional development, and measures to control foreign ownership, among others. The length and incoherence of the policy objectives implied is ample evidence of the absence of a single energy issue or a unified energy policy. Governmental jurisdictions at all levels are involved in regulating, operating and encouraging energy enterprises, each for its own reason. This does not mean that, in the aggregate, total policy will be contradictory: the whole energy sector is so large and complex that various policies can safely modify or bias the prices or quantities of its component parts without necessarily impeding total development.

Indeed, it can be argued that the pervasive intervention of governments in energy matters, which stems from a long list of relatively minor preoccupations, has done little to alter the long-run trends in energy use. While various policies may have hastened or delayed certain developments and significantly altered the distribution of production and incomes, the long-run choice of fuels may not be very sensitive to governmental action. Britain's post-war coal policy and nationalization, Europe's centralization of coal mining under the European Coal and Steel Community, the American restrictions on oil imports, France and Italy's nationalization of integrated petroleum firms, Canada's regulation of electricity and hydrocarbon exports, governmental nuclear energy enterprises in several countries, T.V.A., the British power grid, Electricité de France, the Ontario Hydroelectric Commission – each was significant and important in its time, but in retrospect may have had little lasting effect on the national or international energy picture.

To substantially change the long-run pattern of energy development and use would appear to require a single-minded national determination to go completely contrary to powerful market forces. We can think of only two significant examples. One is the Russian determination, after World War II, to utilize coal rather than oil resources—a decision that permeated and "distorted" the entire development of the Soviet economy for 25 years. The other is the recent determination on the part of some Arab states to obtain high royalties for oil, even at the cost of delaying sale.[12] But it is difficult to find other examples of policies that significantly alter the long-run pattern of energy resources development.

It is equally difficult to predict whether the future will bring a real energy issue that will produce a determination to change energy utilization as profoundly as that in the Soviet Union and the Arabian states. One possibility is environmentalism. Whereas energy production and consumption in the past was rarely, if ever, successfully dominated by a concern for "society," or even for "nationalism," the modern concern for ecological balance and environmental quality may genuinely reshape the demand and supply of energy resources. Indeed, we have already witnessed important antecedents—as in the British demand for hard coal and coke for "smokeless zones"; the imposition of tall stacks on power houses; the reaction against flooding landscape for power dams; the exhaust-abatement requirements on steam engines and internal combustion engines; and California's recent abolition of coal-burning thermal generating plants. And today's jurisdictional, litigative and demonstrative campaigns against the expanded exploitation, use or transportation of energy forms are almost as pervasive as the energy sector itself. In short, while energy policy has not hitherto reflected a unifying motive for public intervention, the growing concern for environmental and social effects may change this by inducing legislators to see energy policies as a single problem, which might lead to a permanent reshaping of energy utilization and development.

Even as we write, however, (in late summer 1973) it seems increasingly unlikely that environmental protection will be a binding constraint on new energy developments. Suffering the inconveniences and shortages of its "crisis," the United States seems now to regard some of its ecological and pollution fears as luxuries to be indulged when energy resources are in surplus, but not when they are scarce. For example, there is now a determination to proceed quickly with a delivery system for Alaskan oil, and Washington has recently announced its intention to relax pollution-control requirements on electrical generating plants and refineries in order to stimulate expansion. While environmental concerns may have permanently restricted the freedom of oil companies and utilities, there seems a general desire for them to go about their business quickly and effectively. Three years ago, the United States environmental movement attempted also to persuade people that growing energy consumption, like economic growth in general, should be questioned and rejected. But this campaign seems to be fading.

Paradoxically, perhaps, the very crisis that is weakening environmental and social constraints on energy development in the United States, appears to be strengthening the Canadian political determination to design some general energy policy. This determination is based on four elements. First, increased American demands for Canadian oil, gas, and electricity, tending to transmit United States scarcities to Canadian markets, have been met by reinforced Canadian energy-trade instruments, especially the controlled licensing of oil and oil-product exports. Indeed, the New Democratic Party, at least, sees energy marketing and prices as a possible 1974 election issue. Second, the manifestation of this increased demand through foreign corporations has led to greater intervention and changes in regulations and quotas, rather than governmental reliance on the self-interest of business. Thus there is, for the first time in many years, a dissatisfaction with royalties on the explicit ground that wellhead prices (on which government revenues largely depend) are not determined at arms length, or competitively. Third, this greater awareness of the revenue aspects of energy policy has been heightened by the continuing re-pricing that is part of a general inflation of all prices and costs. Fourth, as each government tightens up its controls or pricing policies, it tends to encroach on the jurisdiction of other governments. Alberta's new revenue policies, for example, are seen as a threat both to the Ontario government (acting on behalf of many consumers) and the federal government (concerned with conservation and reserves). And a recent federal suggestion that Quebec be included in the protected market for western crude oil has brought critical reaction from both that province and Alberta. All these trends have generated considerable political pressure for a comprehensive energy policy for Canada.

As a result, the first six months of 1973 have seen a great profileration of government activity, which may indeed lead to a mosaic of revenue decisions, resource leases, and export permissions among the two levels of government, out of which might emerge a single, understood "policy." This proliferation has been based on increasing amounts of data, and has resulted in the publication of three important reports by inquiries in Alberta,

Ontario and Ottawa.[13] The latter, the so-called Green Paper, presents an unusual survey of Canadian opportunities and choices, and promises a later White Paper setting forth federal government proposals. Even if this comes to nothing, the work so far and the impetus it has given to study outside the government cannot help but affect forthcoming export decisions, provincial royalty schedules, and judicial hearings on northern native rights.

F. GOVERNMENTAL INVOLVEMENT IN PETROLEUM PRODUCTION IN CANADA

Under Canadian constitutional arrangements, the provinces were accorded jurisdiction over most natural resources, including most fossil fuels. Moreover, with a few exceptions, they have chosen to retain Crown title to sub-surface resources, leaving the exploitation of them to private enterprises. This set of circumstances has led to elaborate arrangements between the public landlord and the private producers.

PROVINCIAL REGULATION

Alberta, which pioneered the development of today's institutional arrangements in Canada—and still produces more than 70 per cent of the oil and 80 per cent of the natural gas in the country— adapted the main features of systems that had been developed earlier in some American states. Superficial and seismic exploration requires a simple licence which provides the province with the means to maintain surveillance and control over general exploration activity. For intensive exploration on specific areas, a permit is required, which provides some exclusive exploration rights for a limited period on the permit area (the maximum size of which is one township, or 36 square miles). In particular, the permittee not only obtains the right to undertake deep drilling, but also the right to acquire production leases that cover 50 per cent of that permit area (or "reservation"). These leases are in the form of rectangular blocks, typically of 8 or 9 square miles on a grid pattern, and a corridor of a mile or so in width must be left between the blocks selected by the permittee. Once the permittee has selected his leases, the remaining areas become

"Crown reserve," which may be sold by auction through sealed bids at a later date. Thus the leases actually covering a discovered reservoir, which carry the right to extract the sub-surface resources for a specified period, are alienated in large part to the explorer, although a number of other operators typically acquire rights to produce from the same reservoir through the purchase of Crown reserves.

The provincial government shares in the value of the resources through three kinds of charges: initial payments for leases purchased from Crown reserves ("bonuses"), annual rentals payable on leases, and royalties which are levied as a percentage of the wellhead value of the oil or gas when it is extracted.

The province also regulates the extraction process to prevent waste of the sub-surface oil and gas and of capital involved in producing them. In particular, wells are subject to minimum spacing rules and maximum rates of withdrawal are specified. In Alberta, these regulations are imposed by the Alberta Energy Resources Conservation Board, a more-or-less independent agency with its own financing from royalties. Other provinces, and the federal government in the northern territories, have adopted modifications of this Alberta system, although they regulate directly through government departments rather than an independent agency.[14]

The need for government intervention in the extractive process to prevent waste and inefficiency arises almost entirely from the fact that the alienation process creates fragmented rights to extract from a common pool. This fragmentation creates incentives for each lessee to produce very quickly in order to obtain as large a share of ultimate reserves as possible. In the absence of restrictions, operators would multiply their wells and extractive capacity and draw up resources rapidly, with the result that reserves would be wasted through dissipation of natural underground pressures, and labour and capital would be wasted in excessive capacity. This problem is usually avoided in the large concessions in the Middle East and Venezuela where private rights to common pools are not fragmented.

With a system that provides incentives to produce at maximum and wasteful rates, the government can attempt to supplement regulations on well spacing and withdrawal rates with more general measures. "Unitization" refers to arrangements for cooperative extraction, and sometimes lessees are required to unitize as a means of achieving more efficient aggregate results. Unitization is often difficult to arrange, however, because it requires agreement on the shares of total recovery attributable to each lease. Moreover, unitized extraction will not prevent the waste in excessive capacity if wells must be drilled before unitization is arranged.

Pro-rationing of the total output required among wells is another measure which Alberta has adopted, copying and improving upon a system developed earlier in Texas. The total demand for Alberta production for each month is determined by the sum of the "nominations" or bids for Alberta oil by refiners. This total is then pro-rated among the pools and wells of the province according to a formula which provides each pool a share based on its ultimate reserves, and each well a share of the pool's allowance according to the area it drains (subject to a guaranteed minimum). By this means, Alberta, which produces most of the country's oil, has traditionally adjusted its output to changes in total demand for crude oil, while the other producing areas operate at capacity.

The systems of alienation and regulation in other provinces differ in detail from that in Alberta. But apart from the system of alienating rights to natural gas in British Columbia, all have a relatively minor quantitative impact on total Canadian production or development incentives. The outstanding exception is in the federal territories of the north and offshore regions. There the system has been adapted to provide considerably stronger incentives for exploration and development.

FEDERAL INVOLVEMENT

The role of the federal government in energy matters has three bases. In the first place, it has over-riding responsibilities for interprovincial trade and commerce, and for external trade. Secondly, it is (subject to possible revision in favour of Indians and Eskimos in the north, and in favour of coastal provinces in offshore energy regions) the owner of "frontier" mineral rights. And thirdly, it is chiefly responsible for the special fiscal treatment offered to the petroleum industry. Taken together, the first two powers mean that the federal government controls the flow of energy from producing regions to other regions and to the United States, and regulates production within the frontier regions. It has less influence on the flow of energy to consumers within the same region (gas from the Peace River to Vancouver, for example, or from Lake Erie to Toronto) and still less on the conditions of exploration and production within the provinces.

The federal landlord. The increasing importance of frontier production has unexpectedly magnified the federal government's direct involvement in regulating exploration and production. As the Ontario Advisory Committee on Energy has recently remarked:

> There is little doubt that over the next five years there will be a substantial change in federal-provincial energy relationships. This is due partly to the changing balance of power in terms of energy supply which also reflects changing geographical emphasis. For the past two decades the Province of Alberta has dominated Canadian hydrocarbon energy supply. Within the next few years the centre of gravity is going to be diffused and the frontier areas will detract from the predominant role which Alberta has played. This will greatly increase the energy policy role of the federal government in hydrocarbon supply.

In its next paragraph, the Committee continues:

> The balance between federal and provincial energy policies is likely to move toward the federal government on yet another count, that is, the growing complexities of the international aspects of Canadian energy policy, including relations with the United States. Canada as a whole and hence the federal government will inevitably become more heavily involved with United States energy attitudes and decisions. If this is so, then more weight will swing towards Ottawa in energy matters."[15]

The "frontier" role, in which the federal government is essentially the landlord of the petroleum industry, is exercised by two bureaux. The energy sector of the Department of Energy, Mines and Resources, is charged with administering offshore mineral resources and other federal programmes and policies apart from those in the northern territories. And the Northern Development Program of the Department of Indian Affairs and Northern Development is especially responsible for the administration and regulation of exploration and oilfield development north of the 60th parallel.[16]

The arrangements that the federal government has made for exploration and development in the areas under its direct jurisdiction provide much more generous incentives than in the provinces, at the cost of great potential loss of Crown control and revenue. The system, explained in detail by Andrew Thompson and Michael Crommelin in Chapter 5, dates from the late 1950's, but most activity under it occurred in the wake of the Prudhoe Bay discoveries in Alaska. Its generous provisions probably reflect the anxious efforts of the Diefenbaker government to lure private industry into the north.

International regulation. The international aspects of federal policy are the particular concern of the National Energy Board. Established in 1959, the five-man Board has statutory power to exercise federal jurisdiction in almost every energy issue.

In fact, however, its activities have been limited to providing information and advice, managing the National Oil Policy and licencing exports and imports of energy.

The Board's advisory and research activities are broad, ranging from the preparation of forecasts and drafting of standards to the inspection of pipelines and the auditing of tariffs and tolls. The National Oil Policy was designed to promote Canadian oil production, at a time of chronic over-capacity in Alberta. It did this be reserving all the Canadian oil market west of the Ottawa Valley for Alberta producers. Low-cost oil from the Middle East and Venezuela was allowed access only to the markets of Montreal and the Maritime provinces. In return for this market protection, the major Alberta producers (most of which are the same companies that supply oil to United States markets from elsewhere) arranged to open the Detroit and mid-western United States markets to western Canadian crude. In September 1973 the federal government proposed that the Montreal market be opened to western crude, but at the time of writing this, this development remains uncertain.

The National Energy Board's authority to licence exports and imports of energy has applied to electricity and gas but not to oil (for which the National Oil Policy has been designed). Since February 1973, however, the Minister of Energy, Mines and Resources has been announcing measures by which Canada may licence the export of oil and oil products. The Board is enabled to intervene in international or interprovincial trade in gas by use of (i) its power to fix rates, tolls and tariffs of pipelines; and (ii) its responsibility to assure that the only gas to be exported shall be proven reserves that are "surplus" to the requirements of Canadian users (after making allowance for further reserves being discovered in the future). The Energy Board Act also directs that the Board shall assure itself that the price at which an applicant proposes to sell gas to an American importer shall be "just and reasonable." This latter duty makes the National Energy Board an offset to the American Federal Power Commission, which is charged with ascertaining that imports from Canada are, from the American point of view, fair and justifiable.

Indeed, in our opinion, the *main* justification for the National Energy Board's existence for its first ten years has been the need to equalise the "national" social presence at the international bargaining table. If the United States Federal Power Commission was attempting to attach stringent conditions to American import licences, Canada had best see to it that the exporter, too, had certain conditions that must be met.[17] More recently, the N.E.B. has been less concerned with the details of a proposed gas-export bargain than with the question of whether Canadian reserves were adequate to permit the export at all. Its interpretation of this responsibility is reviewed in the paper by Paul Bradley in Chapter 3.

Taxation. The third role of the federal government, that of tax collector, appears to be quite separate from that of the landlord and the regulator. In addition to royalties and other revenues from "frontier" petroleum resources, the government is able, by modification of its general income and corporation taxes, greatly to influence exploration and development of oil and gas. For the most part, it has chosen to exercise its influence by manipulating incentives through the tax system. As the 1969 White Paper on taxation explained:

> It is recognized that the exploration for and development of mines and oil and gas deposits involve more than the usual industrial risks and the scale of these risks is quite uncertain in most cases. Consequently, special arrangements are desirable to ensure that the costs of exploration and development may be charged for tax purposes as early as possible in order that taxes will only be applied when it is clear that a project will be profitable. Secondly, it is recognized that the exploration for and development of mineral deposits continue to provide special benefits to Canada and to various provinces . . . [18]

The chief special arrangements in the tax law are: (i) "expensing" of outlays in exploration and development enabling them to be used, dollar for dollar, to reduce taxable income; and (ii) a system of percentage depletion allowances. The latter are changing; by 1977 they will amount to permission to reduce taxable income at the rate of $1 for every $3 spent (and already "expensed" under (i) above at some time in the past) on exploration and development.

These two generous concessions apply not only to Canadian exploration and development but also to foreign operations and to the acquisition of Canadian property for mining purposes. Recent studies suggest that such allowances are excessive and wasteful in the sense that smaller allowances would provide equal incentives to promote exploration, and they lead to misallocation of resources through overinvestment in the extractive industries.[19] Spokesmen such as Eric Kierans have argued that these special incentives have not only distorted the structure of the Canadian economy toward capital intensive activities at the expense of labour employment, but also aggravated instability and contributed "more than any other single policy to the concentration of American ownership that exists in Canada."[20]

Pipeline regulation. The products of the first Canadian oil fields in Ontario and in the Turner Valley were moved to consumers by rail. And most Canadian refiners, whether they used domestic or foreign crude, depended on a combination of tankers and rail cars to keep them supplied. The pipeline was simply a gatherer that moved crude from the well to the terminal or local refinery.

Since the 1950's, however, a network of pipelines has supplanted rail and highway as the means of bringing Alberta oil and gas to western and southern regions. These include not only the gathering systems in and between the oil fields, and trunk lines stretching hundreds of miles to distant market-oriented refineries, but also gas transmission and distribution systems, and lines for carrying refined petroleum products and LPGs. About 70 per cent of the world's pipeline mileage is now used to carry gas.

While the costs of tanker and rail transportation are fairly evenly divided between initial and operating outlays, pipeline costs arise mostly from initial capital expenditure, and are roughly proportional to the length of the line. The cost of transmitting gas, per unit of energy, is at least twice as high as the cost of moving oil. The capacity of pipelines is surprisingly flexible, but operating costs (for pumping oil or compressing gas) rise steeply as the rate of flow is increased. Additional through-put requires larger pipes and "looped" lines. With pipes of larger diameter, the initial cost per unit of daily flow of oil or gas is significantly lower, and so larger flows generally mean lower per unit transport costs. Because of these economies of scale in pipeline transmission, oil companies that compete for reserves and markets nevertheless often cooperate in building pipelines. Thus the 700-mile 24-inch Trans-Mountain Pipeline from Edmonton to Vancouver and Anacortes, and the 2000-mile 16-inch Edmonton-Sarnia-Buffalo Interprovincial Pipeline are owned by consortia of leading oil companies. (Both are instruments of the National Oil Policy, delivering crude to northern United States refineries remote from American oilfields).

Crude oil pipelines are essentially carriers, the oil belonging to the buyer. In Canada, the buyer is usually a refining and market-ing company, which buys crude in Alberta from the various well-owners among whom his demand has been prorated.

There are two main gas transmission lines: the 2500-mile Trans-Canada Pipeline and the 600-mile West Coast system. These are not carriers but utilities – they buy gas under long-term agreements at the wellhead, process it to remove water, sulfur and other hydrocarbons, and deliver it to local gas utilities such as British Columbia Hydro and Consumers Gas in Canada and to allied transmission lines such as El Paso in the United States.

The National Energy Board must approve the construction of all pipelines. Obtaining government or Board approval has always been an emotion-fraught proceeding (as with government involvement in railway planning in an earlier century), and in the famous Trans-Canada case was probably responsible for the fall of the St. Laurent-Howe-Harris government. Furthermore, the N.E.B. must approve the *tolls* on oil lines, the *prices* charged by gas transmission lines, and the export *licenses* of oil exporters and transmission lines, respectively. Gas export, as in the west coast case, has also been a highly-controversial issue. It is diffi-cult to generalize about the issues in these proceedings, but they have had it in common that the promoters have wished to spread their costs and increase their throughputs by building lines to the United States and charging prices low enough to broaden the market there; opponents and intervenors have wished to keep the routes well within Canadian borders or to keep prices to foreign markets high. These same issues have arisen in the con-troversy over the Mackenzie Valley gas pipeline proposal.

G. MACKENZIE DELTA GAS

The current interest in the natural gas resources of the Mackenzie Delta has followed from recent discoveries of oil and gas in
the north slope of Alaska. The existence of oil in Alaska has
been known for many years. Exploration began about 1923, and
the Umiat field in the north slope, which resulted from war-time
exploration, was abandoned undeveloped in 1953 when attention
was swinging to the more accessible Kenai oil field in southern
Alaska. The success of the Kenai stimulated further interest in
the north slope, and in 1967 Richfield (now part of Arco) and
Jersey Standard discovered Prudhoe Bay. Initial indications suggested up to 10 billion barrels of oil-in-place. Other companies
quickly obtained shares in the new field, and a Trans-Alaska
pipeline was planned by a group of companies to carry the oil to
the icefree port of Valdez on the south coast.[21] The Prudhoe Bay
oil reserves contain also natural gas. The State of Alaska has
prohibited the burning off ("flaring") of this gas, and so whether
economic or not, it will be delivered to markets in the cotermi-
nus United States.

These apparently rich discoveries in Alaska triggered new
interest in adjacent areas of the Canadian north. Oil had been
produced at Norman Wells, in the Northwest Territories, since
1921 although the field was small. In the 1960's exploration in
northern Alberta moved into the British Columbia-Yukon border area and discovered gas that was sold to Westcoast Transmission.

Exploration spending in the high Arctic ran to $15 to $20
million per year in the early 1960's. After Prudhoe Bay was
discovered, it jumped to almost $60 million, and one result was
an exceedingly promising oil discovery at Point Atkinson in the
Mackenzie Delta in 1970. Like Prudhoe Bay, this oil reservoir
appeared to be large and free-flowing. Most of the other discoveries in the Delta and in the Melville and King Christian Islands
have been natural gas. The major interests in the Delta are
Imperial Oil, Gulf Oil and Mobil. In the arctic islands, a major
interest in Pan-Arctic, 45 per cent owned by the federal government, which has obtained partial rights in many permits through
"farm-in agreements" and other arrangements. No one yet
knows, however, how oil or gas would be moved from the Arctic
Islands to southern markets.

The immediate interest in the Mackenzie Delta is in the large discoveries of natural gas, not associated with oil (further exploration may reveal oil below the gas, however). Potential reserves of gas in the region are estimated at some 120 t.c.f. Within a few months, a Consortium of companies is expected to apply for permission to proceed with a proposal for developing Mackenzie gas reserves, constructing an enormous 2500-mile, 48-inch diameter pipeline to carry this gas, along with gas produced in the course of oil extraction at Prudhoe Bay in Alaska, to markets in the south. The pipeline alone would cost at least $5 billion. Once installed, it would deliver 4 billion cubic feet of natural gas per day – about one third more than Canada's current consumption – and its capacity could be further increased.

The Consortium, comprised of 26 participating companies, is required to file its application with Canada's National Energy Board and Department of Indian and Northern Affairs, and with the United States Federal Power Commission and Department of the Interior. The project, described in Earle Gray's paper in this volume, raises most of the sensitive economic and political issues that we have mentioned above: the questions of energy trade with the United States, regional development, federal-provincial relations, foreign ownership, competitiveness of secondary industry, disruption of northern native people, and disturbance to the natural environment of the north. The Consortium has already spent some $30 million on feasibility and related studies, and the federal government has also undertaken extensive research on the project, so that a good deal of information on the effects of the project is beginning to emerge. Once the application is filed later this year, extensive public debate can be expected. The question facing Canadians is whether it is in Canada's interest to proceed with the project as proposed, or in some other form, or at a different time.

Construction of the proposed pipeline would be an imense undertaking. Pipelaying would be carried out from 10 large construction camps and two major river crossing crews. In the northern section, construction would be limited to a winter season of almost four months and would incorporate a good deal of new pipeline technology. The optimistic estimates of the Consortium, described in the next Chapter, involve approval of the project within 18 months of application, two years for construction of the pipeline to the Mackenzie Delta, another year for extensions to Prudhoe Bay and first deliveries of gas in the spring of 1978.

Over the 30-year planning life of the pipeline, about half the gas fed into the pipeline is expected to be drawn from the Mackenzie Delta, and half from Alaska, with a larger Canadian proportion during the early years. Most of the throughput would initially be sold in United States markets. Thus the project offers Canada an opportunity to develop and export very substantial amounts of arctic natural gas that would otherwise not be developed for some years.

H. SOME POLICY CONSIDERATIONS

The other papers in this volume discuss the important issues that must be considered in deciding what Canada should do about her new opportunities for developing northern energy resources and the options available to us. Here, it is appropriate to emphasize some general considerations that should be borne in mind in arriving at any decision.

First, the potential gains to Canadians from development and production are theoretically calculable, and although the calculation is frustrated by a paucity of data and uncertainty about the future, best estimates of benefits and costs should be a fundamental consideration. Later papers attempt to throw some light on this question, and the final paper offers a systematic assessment of the gains that could be expected from alternative developments.

Second, we are considering a problem of "marginal" additions to Canada's energy supplies, and not the "total" problem. This simplifies the issue: we can consider the best use of incremental additions to energy supplies from the north without pondering how Canada is going to light, heat and power herself *in toto* over the next fifty years. It can be assumed that because other sources of energy will be forthcoming and bought and sold in world markets, Canada's energy prices and manner of use need not be vastly different from other nations'. Any difference that does emerge will be a yardstick of the cost of our policies. For example, if we decide permanently to dedicate gas reserves to the central Canadian and British Columbia urban markets, the gas prices paid by consumers will be lower than prices abroad by an amount which measures the extra export earnings we are forgoing. Or, if we decide to consume Canadian oil when cheaper oil (or other forms of energy) is available abroad, the increase in prices at the wellhead is a measure of the tax we are levying on ourselves.

But, third, it must be recognized that these benefits and costs will not be equally shared by all Canadians. Particular regions, industries and consumers will inevitably fare well at the expense of others, and will also be influenced by the timing of development. Thus our use of the term "we" in discussing Canada's policy options oversimplifies the choice in omitting the differing impact among Canadians of alternative choices.

Fourth, a decision to postpone the decision is a decision itself, whether it is taken by default or deliberately as a result of well-considered analysis of the benefits of postponement. Certainly we face the opportunity to leave the resources in the ground, and because costs and opportunities constantly change it may be advantageous to develop them at a later date. The cost of postponement is the net benefits sacrificed by not proceeding today, and so the balance of costs and benefits of development at alternative dates is itself a major issue for analysis.

The last two issues are related. Whatever we decide about northern gas development, the availability and cost of energy in Canada is not likely to be greatly changed. Canadians might agree, *as a matter of policy,* to pay slightly more than the world price for certain forms of energy in certain regions or at certain times, perhaps to subsidize fellow countrymen in Alberta or the north. But apart from the deliberate altruism of particular deals we can assume that Canadians will insist on obtaining most of their energy supplies wherever they are chapest, even if this means importing them. Similarly, Canadian consumers may insist on paying less than the world price for certain products through some sort of two-price arrangement. This would amount to a "subsidy" to particular consumers at the expense of foreign consumers and domestic producers. Such a policy can be implemented by means of export taxes or by channelling oil and gas sales through an export marketing board, selling in the United States at higher prices than in Canada. (In the absence of such discriminating policy, Canadian producers would tend to switch supplies to the world market until the price differential was eliminated.)

Thus, although Canada will use different energy sources in different proportions than other countries, their average price and the relative prices among them will differ significantly from world prices only to the extent that we manipulate energy prices for the purpose of redistributing real income. In any event, the issue of export and domestic pricing policy is not likely to hinge on the Mackenzie pipeline question. If we recognize, then, that the decision on northern gas will not significantly alter the structure of the world and Canadian prices for energy, we can examine more readily the question of the terms – when, how and for whom – under which development of arctic natural gas will best serve Canada's national interest.

FOOTNOTES

1. From The National Petroleum Council, *U.S. Energy Outlook,* Vol. II, as cited in Advisory Committee on Energy, *Energy in Ontario: The Outlook and Policy Implications* (Toronto, 1973), p. 20.
2. This problem has been discussed for more than a decade. See Keith C. Brown (ed), *Regulation of the Natural Gas Producing Industry* (Baltimore and London: Johns Hopkins Press, 1972).
3. The National Petroleum Council, *op. cit.*
4. The President of the United States, *Energy Message*, April 1973, p. 3.
5. For a discussion of the probable resolution of the United States energy crisis by a return to diversified self-sufficiency in about 25 years, see E. E. David, Jr., "Energy: A Strategy of Diversity," *Technology Review* (June 1973), pp. 26-31.
6. For a European view, see C. Robinson and E. M. Crook, "Is There a World Energy Crisis?," *National Westminster Bank Quarterly Review* (May 1973), pp. 47-60.
7. In 1965, the United States and Canada showed per capita consumption of energy of 9.7 and 8.1 thousand kilograms of coal equivalent respectively, well ahead of the next highest country, Czechoslovakia, with 5.9 thousand kgs. Most of the statistics in this section are drawn from Joel Darmstadter *et al*, *Energy in the World Economy* (Baltimore: Johns Hopkins Press, 1971).
8. Adapted from *An Energy Policy for Canada: Phase 1*, Vol. I, Department of Energy, Mines and Resources, from Information Canada (Ottawa 1973), p. 73.
9. The seven are (in order of sales) Standard Oil of New Jersey, Royal Dutch Shell, Mobil, Texas Oil (Texaco), Gulf Oil, Standard Oil of California and British Petroleum. Shell is owned by Dutch and British interests, and British Petroleum is British and half government controlled.
10. *Report of the Corporation and Labour Union Returns Act* (1969), Table 2, pp. 126-137.
11. Nor is it an industry which shows significant growth in productivity: the growth in output is attributable almost entirely to increased inputs of labour and capital. See John Dawson, "Productivity Change in Canada in Mining Industries," *Economic Council of Canada Staff Study* (No. 30, 1971), pp. 34-36.

12. Not all Moslem states have followed this policy. Indonesia tried it for ten years, The present position of Iran, Iraq and Libya suggests that the traditional Moslem refusal to admit the existence of a rate of interest stems from a genuine low social rate of time preference.

13. Energy Resources Conservation Board, *The Field Price of Natural Gas,* (Calgary, 1972); The Advisory Committee on Energy, *Energy in Ontario, The Outlook and Policy Implications,* (Toronto, 1972); and Department of Energy, Mines and Resources, *An Energy Policy for Canada, Phase I* (Ottawa, 1973).

14. For a detailed account of the Alberta system, see G. C. Watkins, "Pro-ration and the Economics of Oil Reservoir Development, Province of Alberta, Canada," Ph.D. thesis, University of Leeds, England. (mimeo). The federal system in Canada's Yukon and Northwest Territories is described in Chapter 5 of this volume.

15. Ontario Advisory Committee on Energy, *op. cit.* Vol. 1, p. 28.

16. For references and a general description, see A. R. Thompson, "Sovereignty and Natural Resources—A Study of the Canadian Petroleum Legislation," *UBC Law Review,* Volume 4, No. 2, p. 161; and Chapter 5 in this series.

17. Paul Bradley's paper in Chapter 3 discusses gas export licencing in detail. For a short description of an Energy Board ruling on prices, see Ralph S. Spritzer, "Changing Elements in the Natural Gas Picture," in K. C. Brown (ed) *Regulation of the Natural Gas Producing Industry* (Baltimore: Johns Hopkins Press. 1972), pp. 129-31. For a description of pre-N.E.B. negotiations, see H. G. J. Aitken, "The Midwestern Case: Canadian Gas and the FPC," *Canadian Journal of Economics and Political Science,* Vol. 25 No. 2, (May 1959), pp. 129-43; and "The Changing Structure of the Canadian Economy in B. U. Ratchford (ed), *The American Economic Impact on Canada*(Durham, N.C., 1959), pp. 3-35.

18. Hon. E. J. Benson, *Proposals for Tax Reform* (White Paper on Taxation), Ottawa, 1969, p. 64.

19. *Report of the Royal Commission on Taxation* (Carter Report), Ottawa, 1966, Volume 4 Chapter 23. The new depletion provisions, requiring taxpaying corporations to "earn" percentage depletion allowances at the rate of $3 of exploration and development spending for every dollar of depletion, are partly in response to the Carter Commission's criticisms of the previous system, and hence may prove to be less wasteful of tax concession dollars in achieving the aims cited on p. 64 of the White Paper (quoted earlier.)

20. Hon. Eric Kierans, P.C., M.P., "Contribution of the Tax System to Canada's Unemployment and Ownership Problems" Address to the Annual Meeting of the Canadian Economics Association, St. Johns, 1971 (Mimeo).

21. See C. J. Cicchetti, *Alaskan Oil: Alternative Routes and Markets,*Resources for the Future Inc. (Washington, D.C., 1972).

2.
WHY CANADA NEEDS
THE ARCTIC GAS PIPELINE

Earle Gray

Canadian Arctic Gas Study Limited and our affiliate, Alaskan Arctic Gas Study Company, are engaged in detailed examinations of the engineering, economic and environmental aspects of a proposed $5 billion pipeline to transport natural gas from the Mackenzie Delta and the North Slope of Alaska. Twenty-eight firms are participating in these studies, representing a major segment of the petroleum and natural gas industries in Canada and the United States. Eleven of these firms are majority owned and controlled in Canada; two are controlled overseas, and the remainder are United States controlled firms. We have executive offices in Toronto with engineering offices in Calgary, and a present staff approaching 100. In addition, some 30 consulting organizations are engaged in our study programs.

Alaskan Arctic Gas Study Company has established offices in Anchorage. It will supervise study programs conducted in that state, and will also provide information to, and consult with, government and public sectors concerned with the transportation routing for shipment of Alaskan gas.

Arctic Gas had, by mid-1973, spent more than $35 million

collecting the information required to examine the proposed pipeline and assess the implications. It is the intention of Arctic Gas to file with Canadian and United States government agencies as soon as we have obtained all of the information that we feel is required to properly support applications for leave to construct and operate the proposed facilities. No other project will have been subject to more extensive and thorough research and study. And no other project will be subject to more exhaustive examination at public hearings before government agencies than those hearings that will follow the filing of our applications.

An application for a certificate of public convenience and necessity will be filed with the National Energy Board, which would make a recommendation to the Federal cabinet. Application will also be made to the Department of Indian and Northern Affairs to obtain the use of federal lands for a right-of-way across the Yukon and Northwest Territories.

DINA has announced it will hold public hearings on this application, including hearings in the Territories, in order to provide northern residents opportunity to present their views. In the United States, application will be made to the Federal Power Commission and the Department of the Interior in respect to the facilities in Alaska, and the importation of gas into the United States.

It is apparent that the government of Canada could approve this pipeline only if thorough consideration of all of the relevant aspects clearly demonstrates it to be in the national public interest. It is also apparent that considerable time will be required for the regulatory proceedings and public hearings before a final determination from the government can be expected. If it requires, say, 18 months from filings to final approvals, Arctic Gas will by that time have spent some seven years and $70 million on the feasibility studies, regulatory proceedings, and detailed pre-construction engineering.

In financing the pipeline, first priority and opportunity will be given to Canadian investors, including corporate, institutional and private investors. The only limitation to Canadian ownership will be the amount of money that Canadians may choose to invest in this pipeline. We are confident that, under favorable market conditions, this will exceed 50 per cent.

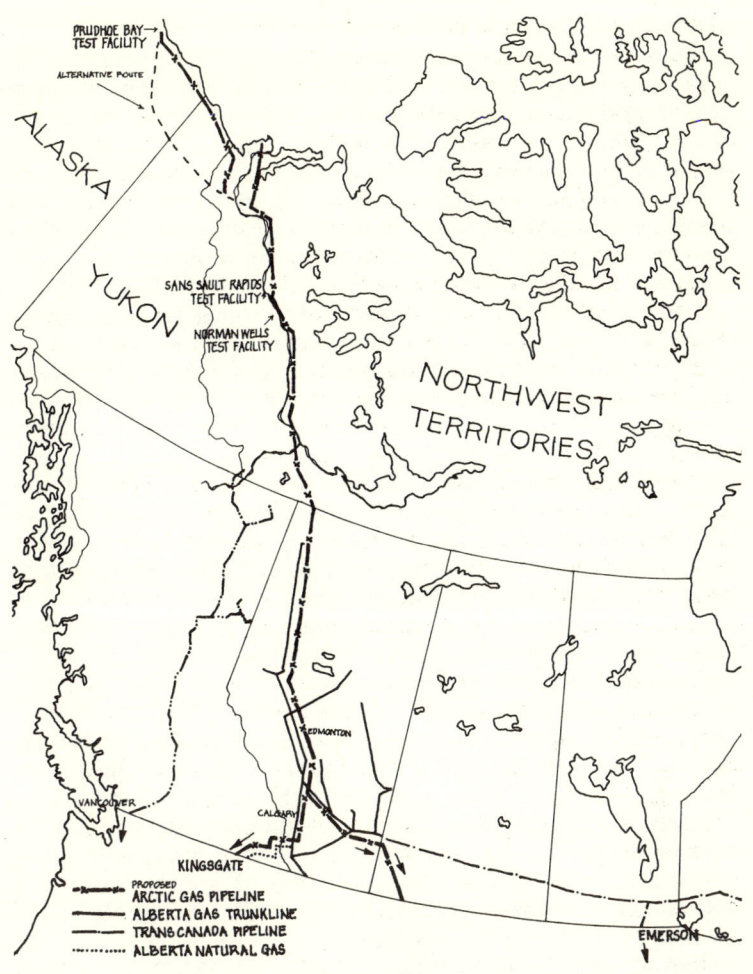

PRUDHOE BAY
TEST FACILITY

ALTERNATIVE ROUTE

ALASKA

YUKON

SANS SAULT RAPIDS
TEST FACILITY

NORMAN WELLS
TEST FACILITY

NORTHWEST
TERRITORIES

EDMONTON

VANCOUVER

CALGARY

KINGSGATE

PROPOSED
ARCTIC GAS PIPELINE
ALBERTA GAS TRUNKLINE
TRANS CANADA PIPELINE
ALBERTA NATURAL GAS

EMERSON

A. PROPOSED PIPELINE FACILITIES

The proposed pipeline facilities will comprise approximately 1,900 miles of 48-inch diameter pipeline, and 650 miles of 42-inch diameter pipeline. Of this, in the order of 200 miles would be in Alaska, and the remainder in Canada. The proposed route is shown on the accompanying map. Forty-eight inch diameter pipe extends from the gas supply areas, up the Mackenzie Valley, and through Alberta to a point northwest of Calgary at Caroline. At this point the system splits, with one 42-inch diameter leg extending southwest, and the other 42-inch diameter leg extending east. All of the northern gas for Canadian markets east of Alberta will be picked up from the Arctic Gas pipeline on the Alberta-Saskatchewan border at Empress by TransCanada Pipelines. In turn, TransCanada would move this gas across Saskatchewan, Manitoba and Ontario. This would assure sustained construction expansion on the TransCanada pipeline across the prairie provinces and Alberta for a long period. It will also provide protection for the existing Alberta Gas Trunk Line and TransCanada facilities against the day of declining deliveries from the gas supplies of western Canada.

In the far North, alternative routes are still under consideration. They involve two principal options from Prudhoe Bay across Alaska and the northern Yukon. One route lies east from Prudhoe Bay along the coastal plain, then south up the Mackenzie River Valley. This route would cross the Alaska Wildlife Range. The alternative is to skirt the Wildlife Range by crossing the Brooks Mountains in Alaska and the Richardson Mountains in the Yukon. The coastal route is shorter, appreciably less costly, would traverse a greater area of potential gas supply, and, we are advised by our consultants, would involve less risk of environmental impact. The pipeline will be a contract carrier. Unlike most natural gas pipelines, it will neither buy nor sell gas, but will simply provide transportation for shippers of gas who contract for this service. It will be essentially a utility-type operation with a regulated rate of return.

With installation of full compressor horsepower, the pipeline would be able to deliver in excess of four billion cubic feet of gas per day. Additional capacity can be installed as required by means of incremental looping with more large-diameter pipe.

The unique design feature of the proposed pipeline involves the refrigeration of the gas. Throughout the areas of permafrost, roughly north of 60 degrees latitude, the gas will be refrigerated at each compressor station to a temperature of about 25 degrees Fahrenheit, so that it will be close to the soil temperature, and below freezing. This chilled gas concept will allow the pipeline to be fully buried along the entire length, and avoids the risk of any permanent damage to high ice-content permafrost. The surface over the buried pipe will be revegetated to provide a thermal insulation for the permafrost, and to arrest alluvial erosion. The right-of-way would show little evidence of a pipeline. There would be a levelled, narrow clearing through the brush of the tioga, which would afford no impediment to the wildlife, nor to surface water drainage. In the tundra, there will be little visual evidence of even the right-of-way. The most noticeable evidence of the pipeline will be at the compressor stations every 50 miles or so, and possibly at a very few elevated river crossings. The right-of-way and compressor station sites will occupy about 40 square miles in the Yukon and Northwest Territories, out of an area in excess of 1.5 million square miles.

Allowing for the regulatory procedures plus assembly and delivery of materials, it is unlikely that construction work could start prior to the winter of 1976-77. Construction in the North would be limited to a winter season of about four months. Equipment would move over snow and ice roads to protect the vegetative mat which insulates the permafrost. With careful planning it may be possible to build the northern end of the system as far as the Mackenzie Delta during two winter seasons, with the extension to the Alaskan North Slope during the following winter. This would mean that gas might start flowing from the Delta by mid-1978 and from Alaska by mid-1979, assuming that there are no delays.

B. STUDY PROGRAMS

The study programs being conducted by Arctic Gas embrace aspects as diverse as metallurgy, soils mechanics, thermodynamics, economics, wildlife biology, and sociology, to mention only a few. It is difficult to imagine a research program involving a greater range of disciplines. Principal study programs include the following: Pilot test pipelines at Sans Sault and Norman Wells, Northwest Territories; and Prudhoe Bay, Alaska. Chilled and compressed air used to simulate the flow of natural gas has been circulated through 5,000 feet of 48-inch diameter test pipes. In operation for nearly two years, these test facilities have demonstrated that fully-buried, chilled gas pipelines can be constructed and operated without any permanent damage to the permafrost. This $7 million program has provided data on the following: stability of gas pipelines in permafrost under various simulated operating conditions; stability of different types of foundations for above-ground structures; effects of a pipeline on various forms of surface cover; drainage problems associated with Arctic pipeline construction; and construction methods, design techniques, material and equipment for use in an ultimate pipeline system.

Environmental studies, which had cost $8 million by mid-1973, include both wildlife and vegetative studies in northern Canada and Alaska. The wildlife studies have defined possible areas where construction and operation of the pipeline might conflict with the wildlife resources, and means to avoid or minimize such conflict. The vegetation studies have shown how to restore a vegetation cover over a buried pipeline in order to minimize thermal erosion of the underlying permafrost in the vicinity of the pipeline, and to arrest alluvial erosion. Base line studies completed to the end of 1972 provided an immense amount of valuable information on the habitat, populations and characteristics of wildlife resources through a large segment of the North. Several million dollars more are being spent on environmental studies which now focus on specific impact areas. They show, for example, such factors as the timing and location required for river crossings to avoid disturbance of fish; other critical periods and locations where construction activity may have to be restricted to avoid disturbance to caribou when they are calving, or migratory birds during their brief pre-migration feeding period on the Arctic coastal plain. Some of the other areas of study underway by Arctic Gas include the following: (1) Transportation alternatives, to provide an assessment of the engineering, economic and environmental factors involved in alternate modes and routes for transporting natural gas, such as railroad. (2) Economic studies, to assess the national and regional impacts of financing and operation of the pipeline on the Canadian economy, including such aspects as employment, balance of payments, foreign exchange rates, and interest rates. (3) Financing studies. Our objective is a viable plan of financing which will provide first priority to Canadian investors with more than 51 percent Canadian equity ownership. (4) Regional socio-economic impact of the pipeline in the Northwest Territories and Yukon. (5) Recruitment, training and employment of northern residents for the construction and operation of the pipeline. A program employing northern residents on training programs on the operating facilities of Alberta Gas Trunk Line has been underway for Arctic Gas in Alberta for more than two years. Results have been most encouraging.

Successful trainees are offered permanent employment with Alberta Gas Trunk Line, and will have an opportunity to work on the operation of the proposed northern pipeline. Working with both federal and territorial government agencies, and with other companies in the petroleum industry, we are now involved in more comprehensive training programs which embrace not only pipeline operation, but also exploration, drilling, producing and gas processing plant operations in the North. More than 70 northern residents were participating in these programs by late 1973.

C. RISKS OF DEFERRAL

Canada cannot afford a lengthy deferral of a pipeline to transport natural gas from the Mackenzie Delta, for several reasons: (1) Canadian consumers will, within a relatively few years, require supplies from this northern region if shortages are to be avoided. (2) The transportation of gas from the North Slope of Alaska by pipeline across Canada to United States markets would help provide the scale of operations necessary for the economic transportation of Canada's northern gas resources. If the North Slope gas were to be moved to United States markets by an alternative trans-Alaska route, this would seriously impede and delay the availability of Canada's northern gas resources.

We are now dependent almost exclusively on the sedimentary basin in the four provinces of western Canada for our entire supplies of natural gas. Yet Western Canada accounts for less than 15 per cent of our total estimated potential gas resources, as shown on the following table.

Of the estimated 113 tcf potential gas resources in Western Canada, some 69 tcf, or 60 percent, had been discovered by 1973. The second half of these potential gas resources in Western Canada will be increasingly difficult and expensive to discover. Reserves of natural gas in Western Canada are not now growing at a rate as fast as the growth in demand. The result has been a

Canada's Proved and Potential
Gas Resources

Region	Total, proved & protential gas, trillions of cubic ft.	% of Total
Western Canada	112.9	14.4
Northern Canada, ex Islands	101.0	12.9
Arctic Islands	240.7	30.7
East Coast	326.1	41.7
Other	2.2	0.3
Total	782.9	100.0

Source: Geological Survey of Canada

gradual decline in the relative supply of proved remaining reserves of marketable natural gas. This has not yet reached a critical point. But the trend will continue until we can gain economic access to the 85 per cent of our potential gas resources that lie outside of Western Canada. The present and potential gas reserves of Western Canada will not be depleted for probably several decades. But there are physical limitations on the rate at which these reserves can be extracted. As gas is drawn from a field, the reservoir pressure declines and there is a corresponding gradual decline in the rate at which the gas can be produced. This means that, long before we have depleted our gas reserves in Western Canada, the rate at which we can draw from them will not be sufficient to supply continually increasing demand. Studies by Arctic Gas indicate that by 1980 the supply of natural gas available from present and anticipated reserves in western Canada will not be sufficient to satisfy domestic demand, plus the specified volumes which have already been authorized by the government of Canada for export to the United States. The amount required to supplement production from western Canada will not initially be large, but it will increase steadily as a result of growing domestic demand, combined with a gradual decline in the production rate of western gas.

The volume of supplementary supplies required by Canadian consumers will not be sufficient to provide attractive economies of scale for the transportation of gas from the Mackenzie Delta. However, U.S. markets coupled with U.S. gas from the North Slope of Alaska can provide the demand and supply required to make the project economically viable. As presently envisioned, at least half of the pipeline throughput at full design capacity would be U.S. gas from the North Slope, with the other half from the Canadian North. All of the Alaskan gas would flow to U.S. markets, as would a portion of the Canadian gas, provided that the Canadian gas is found by the federal government to be surplus to anticipated Canadian requirements.

If the Alaskan North Slope gas were not available, a very much greater volume of proved reserves would have to be discovered and developed in the Mackenzie Delta, in order to make the pipeline financible. And a much greater portion of the potential gas resources of northern Canada would have to be exported to the United States in order to attain the required volume of deliveries. The probable result of this would be to defer the availability of Mackenzie Delta gas well beyond the point in time that it will be required to meet Canadian needs. This possibility does exist. An $11 million feasibility study of an alternative transportation route for the North Slope gas is already well underway by El Paso Natural Gas Company. The proposal envisions a pipeline from Prudhoe Bay to Alaska's Pacific Coast where the gas would be liquefied for tanker shipment to California. We do not believe that this is the most economic nor the most advantageous means for the United States to gain access to its North Slope gas reserves. But it is feasible, it is a serious proposal, and if the United States were to face lengthy delay in gaining access to its Alaskan gas by means of a pipeline across Canada, it would have little option but to turn to a trans-Alaskan delivery route, as it seems likely to do in the case of oil.

Approval of the Arctic Gas pipeline will be required without excessive delay in order to (1) Avoid future shortages of natural gas supplies for Canadian consumers. (2) Take advantage of the movement of Alaskan gas across to minimize the cost of Canada's northern gas.

D. ECONOMIC IMPACT

Studies underway by Arctic Gas include the use of computer models to simulate the projected Canadian economy, with and without the pipeline. The purpose is to examine the possible impacts of the pipeline, during both the construction and operating phases, in terms of such factors as employment, national income, balance of payments, dollar exchange value, interest rates, and cost of goods and services. A number of computer projections are being run under varying assumptions in order to test these impacts under a wide range of possible conditions. These studies are not yet completed. A number of preliminary conclusions, however, are indicated.

During the construction phase, the pipeline could make a major contribution to national employment and income, the extent dependent in large measure on the degree of slack in the economy at the time. Potential adverse macro-economic effects are expected to be negligible. Little change in the value of the Canadian dollar is expected to result from the construction, as direct and induced capital flows attributable to the pipeline largely offset one another. Sustained, long-term benefits expected from the operational phase of the pipeline include the following: (1) Assurance of continuity of supply for Canadian industry and other consumers. (2) Direct and indirect employment resulting from the operation of the pipeline and the attendant exploration and producing operations in northern Canada. (3) Generation of appreciable government revenue from Crown-owned oil and gas rights.

It is difficult to over-estimate the economic importance of an assured and adequate indigenous supply of energy. Some indication of this is provided by the concern confronting the United States as a result of the widening short-fall in its domestic energy supplies. Canada is now the only industrialized nation in the world that is self-sufficient in energy.

Critics of the pipeline have generally failed to recognize the full impact on employment that it will produce. The pipeline is but one aspect in the chain of operations required to find, develop, produce, process, and transport Mackenzie Delta gas reserves. Employment creation can be looked at realistically only by considering this total operation, and its requirements for manufactured supplies, services and transportation. A general indication of the total employment impact is provided by the pipelines which transport virtually all of the oil and gas currently produced in Western Canada. These pipelines directly employ only some 3,500. But the total exploration-production-transportation operations provide a total direct and indirect employment in Western Canada of more than 200,000. In addition, there is the employment resulting from the industry's purchases of more than $10 billion worth of Canadian-manufactured goods since 1947. These are not short-term jobs. Nearly 30 years of active exploration and production in the western provinces has resulted in the discovery of about 60 per cent of the estimated potential oil and gas reserves of this region, and production of less than 15 per cent. In similar fashion, the long-term employment associated with a gas pipeline from Alaska and the Mackenzie Delta will amount to many thousands of jobs, not only in the North but throughout the national economy.

Operation of the pipeline and the export of surplus gas will result in a net generation of foreign exchange earnings amounting to a few hundreds of millions of dollars per year. In view of a deficit in the current balance of payments projected by the Economic Council of Canada at an average rate of more than $1.1 billion per year for the 1973-1980 period, these earnings could provide an important contribution to Canada's continued prosperity. Pending promised revision of the Canada Oil and Gas Land Regulations, it is not possible to project the amount of revenue that the government will collect on natural gas production from leases to Crown-owned petroleum rights in the Mackenzie Delta area. Even the former lease provisions (now in abeyance) would have generated several hundreds of millions of dollars in government revenues. Ultimately, there is no reason why oil and gas production in the North should not produce at least as much government revenue as Western Canada. Here, the four provincial governments have to date collected net revenues in excess of $4 billion from oil and gas production on leases to Crown-owned oil and gas rights.

E. NORTHERN INTERESTS

The 50,000 residents of the Northwest Territories and Yukon – especially those in the Mackenzie Valley region – have a particular interest in the pipeline and the development of northern gas resources. There are three principal areas of concern. First, there is the need to protect the natural environment and wildlife resources of the North. Related to this is the desire to maintain unimpeded opportunities for hunting and trapping, whether as full-time pursuits, part-time activities to supplement wage incomes, or as recreational pursuits. Second, there is the need to sustain a viable northern economy with the provision of adequate jobs for northern residents, particularly native peoples. Third, there is the resolution of northern native land claims.

It is clear that construction and operation of the pipeline, and the associated gas development, must be such that it will not result in intolerable impacts to the northern environment nor impede the opportunities for hunting and trapping. Based on our extensive studies, we are wholly confident that in our applications and resulting public hearings it can be satisfactorily demonstrated that this objective can be accomplished. But just as it would be wrong to impede the opportunities for hunting and trapping, surely it would be equally as wrong to deny to those northerners who may seek it, the opportunities for wage employment. Conditional support for construction and operation of a gas pipeline from the Mackenzie Delta has been clearly expressed by northern residents. Last February, the Legislative Council of Northwest Territories moved to "formally recommend and support the construction of a pipeline . . . through the Mackenzie Valley." In May, James Wah-Shee, President of the Federation of Natives North of '60, made the following statement to an international conference: "We believe that the extraction of our northern mineral resources are perhaps in the best interest of the native people and the developers, if the native people are assured that our share of the wealth will allow our people to compete as equals with the rest of society." There are qualifications attached to such support. These include settlement of the native land claims, and ample opportunity for northern residents to benefit from employment and other opportunities resulting from the pipeline. Settlement of northern native land claims is certainly beyond the authority of Arctic Gas. But it is hoped that this can be resolved without impeding the development of northern resources, for the benefit of all Canadians, particularly northern Canadians.

We are convinced that the long-term job opportunities in the North will substantially exceed the available labor force of the region if the pipeline is built. The real challenge will thus be to provide the training programs and other measures which will ensure that residents of the North will be able to take advantage of these opportunities. The training program in which Arctic Gas is participating, along with other members of industry and the territorial and federal governments, is designed to meet this need.

Production of the far northern gas resources is still a number of years in the future. But already the northern economy and employment are being stimulated by the expanding tempo of exploration activities. Last year, for example, oil and gas companies spent $243 million in their search for northern oil and gas, an increase of nearly 40 per cent from 1971. This has already provided jobs for several hundred northerners. Given the prospect of a pipeline and the encouraging exploration results to date, there is every reason for confidence that this level of activity will be not merely sustained, but will be increased. The North has a vital interest in seeing this activity sustained. The present economy and level of employment in the North is based on anticipation of large-scale petroleum and minerals development, according to Mr. Ewan Cotterill, Assistant Commissioner of the Government of the Northwest Territories. In an address last May, Mr. Cotterill told the Conference Board of Canada that if development of the North's resources does not take place, or if it takes place much later than expected, the effect on the Northwest Territories could be disastrous.

APPENDIX TO CHAPTER 2
SUMMARY OF MAJOR FEATURES OF THE
PROPOSED PIPELINE

Capacity: 4.5 billion cubic feet per day, of which 0.4 bcf to power the system.

Cost: Study and application – $60-70 million.
                   Construction – approximately $5 billion invested over a 5-year period.

Schedule: Construction start, winter 1976-77.
(Proposed)   In-service carrying Delta gas, Fall 1978.
                   In-service carrying Delta and Prudhoe gas, Fall/79 or possibly Fall/78.

Mileage: Prudhoe Bay Branch (coastal route)   492 miles (48")
                   Mackenzie Delta Branch   126 miles (48")
                   Mackenzie Valley Line to Caroline, Alta.   1303 miles (48")
                   West Line – Caroline to Kingsgate, B.C.   283 miles (42")
                   East Line – Caroline to Monchy, Sask.   394 miles (42")

                   Total length of pipeline via coastal route – 2598 miles
                   Length of line in Canada – 2403 miles
                   Length of line in Alaska – 195 miles

**Pipe and
Station Data:** Operating pressure – 1680 psi.

Estimated installed compressor station and propane refrigeration station horsepower at full capacity:

Gas compression – 1,690,000
Refrigeration compression – 288,000

**Land, Material
& Manpower
Requirements:** Estimated land requirements for temporary construction, right-of-way and permanent facilities:

Alaska – 7 square miles
Canada, North of '60 – 40 square miles
Canada, South of '60 – 40 square miles

Estimated material requirements for project:
2,012,000 tons 48" pipe
508,000 tons 42" pipe
2,520,000 tons TOTAL mainline pipe

170,000 tons of compressor station equipment
160,000 tons of contractors' construction
equipment and camps
500,000 tons of fuel and methanol

Manpower requirements for construction:
8,000 at peak with 5,000 to 6,000 average.

Manpower requirements for operation:
more than 600 in Canada and 100 in Alska,
excluding head office requirements for both
regions.

3.
ENERGY, PROFITS,
AND THE NATIONAL INTEREST:
THREE PERSPECTIVES
ON ARCTIC NATURAL GAS

Paul G. Bradley

The fundamental premise underlying the proposed arctic natural gas pipeline is elegantly simple: large quantities of gas are available in the barren north of the continent, while large quantities of gas are needed in the populous south, particularly in the United States. This premise is solidly grounded. Specialists declare that the Arctic is destined to become one of the major petroleum-producing provinces of the world.[1] Such optimism about availability comes at a time when a remarkable sequence of events has caused dislocations in world trade patterns in energy resources, as well as changes in the conditions under which these resources, notably petroleum, are supplied. In the United States adjustment to these changes is accompanied by the threat of energy shortages, and this threat has become a matter of national concern.

It is little wonder that in a situation of change and uncertainty, the fear of shortage engenders dire warnings, while the anticipation of potential riches encourages rash claims. School closures in Colorado are bound to raise the spectre of dark, unheated homes. Gas finds in the Arctic encourage dreams of Arabian riches, or, alternatively, a compulsion to hoard the new symbol of wealth. In an atmosphere of crisis, the immediate question in the United States is this: what quantities do we need? In Canada, with an apparent surplus of gas, the corresponding questions are these: what quantities do we have and how much should we hold on to?

Taking a more detailed look at the proposed large-scale development of arctic natural gas, we find that diverse groups have expressed concern. Foreign and Canadian investors have already advanced substantial sums in anticipation of future profits accruing to the export of arctic gas. Eastern gas consumers see decisions about the allocation of natural gas as having a large effect on their costs, and Eastern manufacturers are anxious to be guaranteed preference in the awarding of large contracts for pipe. Environmentalists are alarmed by the possibility of widespread destruction both in the producing area and along the pipeline. Looking to the North, the establishment of petroleum production is viewed variously as the mainspring of bountiful economic development or as the potential scourge of the native people. Thus many Canadians foresee pipeline consequences which will directly affect them. Though not so obvious, large-scale exports of natural gas are likely to have substantial, possibly adverse, effects on the development of the Canadian economy.

If the full implications of the gas pipeline are to be understood, public discussion will have to penetrate beyond the rhetoric of needs and availabilities. There is much analysis to be done, particularly in the economic sphere; other writers in this volume demonstrate what can be accomplished. In this paper we propose to examine the positions staked out by the major participants in Arctic natural gas development. We identify three major parties:

(1) the recipients of the exports – American consumers, whose interests are represented by the United States government;
(2) the intermediaries – private companies who secure title to the gas, produce and transport it, and then sell it; and
(3) the resource owner – the Canadian government, which sets the terms of alienation of the resource and the conditions under which it is produced and transported.

We will attempt to shed some light on the objectives so far revealed by the actions of each player in this three-way game, as well as to point out some of the constraints on their actions.

Two further comments will clarify the scope of this paper. For the Canadian government, the most visible and active agency has been the National Energy Board (NEB). We will examine its policy objectives in some detail and compare them with the customary goals of economic policy. Finally, although this book focuses on a single pipeline proposal, the Mackenzie project is almost certainly a bellwether. A new consortium, Polar Gas Project, has been formed very recently to study underwater crossings from the Arctic Islands to the mainland and possible pipeline routes southward to central and eastern Canada and the United States. Panarctic Oils, Ltd., is well advanced toward establishing sufficient reserves in the Arctic Islands to justify construction of a large-diameter pipeline.[2] Still further to the east, discoveries offshore in the Atlantic foreshadow additional large flows to consuming areas.

A. CANADIAN GAS IN THE UNITED STATES MARKET

In analyzing the market for any particular energy resource, one must constantly bear in mind the interdependence among all energy resources. For example, the natural gas shortage currently being experienced in the United States has been blamed (apparently with considerable justification) upon the Federal Power Commission's policy of enforcing ceilings upon field gas prices.[3] As demand increased, these ceilings prevented the development of higher cost gas fields. However, the rapid rise in the demand for natural gas, particularly for commercial and industrial uses, was itself closely related to federal policy toward another fuel: maintaining an artificially high price for crude oil has been a feature of United States energy policy since the 1950's, and this has favored more widespread use of natural gas.

Estimates of market potential in the United States for Canadian natural gas would ideally be derived through detailed study of United States energy demand and the supplies of competing domestic and foreign fuels. It would be recognized that at any time the demand for a particular fuel is affected by the relative price of other fuels, and that its supply is influenced by past rates of use and discovery. Modeling the energy sector on the premise that markets are competitive would be difficult enough, but the need to account for noncompetitive market behavior and the imposition of government regulation further complicates the task.

TABLE 1
Projections of United States
Natural Gas Supply in 1985
(By source—Tcf/Year)

Case:	I	II	III	IV
Lower 48				
Onshore	17.1	15.2	12.0	9.6
Offshore	9.1	7.8	5.5	3.6
Alaska, North Slope	3.3	2.7	2.2	1.3
Alaska, South	1.1	.9	.6	.4
Total Conventional Domestic Production	30.6	26.5	20.4	15.0
Synthetic Gas (SNG)	3.8	2.6	2.6	1.8
Gas from Nuclear Stimulation	1.3	.8	.8	-
Imports				
Liquified (LNG)	3.2	3.4	3.7	3.9
Pipeline (Canadian)	2.7	2.7	2.7	2.7
Total Supply*	41.6	36.0	30.2	23.4
Total Imports, percentage	14	17	21	28
Canadian Imports, percentage	6.5	7.5	8.9	11.5

* Totals may not agree due to rounding.
Source: National Petroleum Council, *U.S. Energy Outlook, A Summary Report,* Dec. 1972, p. 58.

However, if the provision of energy does turn out to be the chosen vehicle for American technology in this decade – as moon rockets were in the sixties – we can expect a generation of energy-sector models to be propagated. The computers are already humming. In the meantime, it will be instructive to draw on available information to attempt a rough sketch of the position of Canadian gas exports at the time an arctic pipeline could be put into operation.

An indication of the role of Canadian gas in supplying the United States market can be seen in projections recently completed by the National Petroleum Council (NPC).[4] Future energy needs of the United States were derived by extrapolating rates of increase of energy consumption. Availabilities of domestically produced crude oil and natural gas were computed by extrapolating trends in exploratory drilling activity and finding rates, while independent forecasts of the quantities of energy to be derived from other sources (chiefly coal and nuclear fuel) were accepted. Different assumptions – high and low, pessimistic and optimistic – were made for the variables describing need and availability. For each selected combination of assumptions, patterns of fuel use were projected for 1975, 1980, and 1985. Consistency was obtained by working out the material balances; in these balances imported crude oil was treated as the residual, making up the projected gap between energy consumption and energy derived from all other sources (these sources including all domestic fuels from synthetic and natural sources plus natural gas imports, natural gas being the only other imported fuel).

The NPC method does not admit changes in use patterns because of substitutions in response to changed relative prices. In fact, prices – that is, exchange values – do not enter into the analysis at all. Instead, "the term 'price' is used to refer generally to economic levels which would, on the basis of the cases analyzed, support given levels of activity for the particular fuel."[5] Thus, the so-called "prices" are derived *ex post*, taking on the values required to guarantee a specified rate of return on projected investment. The failure to account for price-induced changes in the use of different fuels is a serious deficiency of the method, as E. Berndt emphasises in the next Chapter. Nevertheless, though we must recognize the likelihood of a large margin of error, the NPC figures still provide a starting point for appraising the prospects for Canadian natural gas exports.

Table 1 shows NPC projections of United States natural gas consumption in 1985, with the totals subdivided by source. In all cases the bulk of American consumption is supplied from domestic sources: the highest projected share for imports is 28 per cent. It is in this case that imports from Canada also assume their greatest importance, accounting for between 11 and 12 per cent of the United States market. Should this projection prove accurate, Canadian gas would have gained a significant market share, though certainly not a dominant one. To forecast the price that Canadian exports are likely to command we must explore the economic dimension which is missing in the NPC report.

Future natural gas consumption in the United States will depend on price, with government policy continuing to play a major part in price determination. Supply and demand conditions are, of course, fundamental. Aggregate demand will depend on the price of gas relative to alternative fuels; aggregate supply will reflect the costs of the various sources. In recent years, with Federal Power Commission ceilings on the field price of natural gas, price has not moved up to the market-clearing level. This form of intervention will almost certainly be withdrawn, but other forms can be expected to replace it. The advocates of a competitive, free-enterprise energy sector tend to interpret very generously the government's responsibility for maintaining an economic climate conducive to private enterprise. In particular, the United States government will probably limit imports, offer increased tax incentives to domestic natural resource industries, subsidize the development of new technologies for synthetic fuels, and provide the infrastructure investment needed to exploit unconventional sources (for example, aqueducts to supply the large amounts of water needed in exploiting oil shale or tar sands deposits). With regard to the value of Canadian natural gas in the next decade and beyond, import controls are the most important of these measures.

As explained in Chapter 1, the United States has for more than a decade manipulated the tradeoff between price level and degree of self-sufficiency. This type of choice will continue to be a feature of policy with respect to energy in the aggregate and to particular fuels. The United States still has substantial indigenous reserves of natural gas which can be developed; by guaranteeing a sufficiently high price level to promote the exploitation of more costly natural gas sources, it might even be able to maintain its present level of self-sufficiency. This possibility is enhanced when account is taken of the substitution of alternative fuels as the relative price of natural gas increases. Maintaining a high degree of self-sufficiency would in all likelihood require import controls, since the domestic price would reach a high enough level to make attractive both liquified natural gas (LNG) imports and synthetic gas (SNG) produced from imported liquid hydrocarbons. At this point one can do little better than guess at what quantities of natural gas the United States will import toward the end of this decade and during the next. The NPC figures of Table 1 offer one projection, and we will consider the implications of the situation which they portray.

The NPC foresees United States domestic production being supplemented, aside from Canadian imports, by two principal sources, LNG and SNG. Both of these can be expected to be more responsive to price than Canadian imports. With the construction of large-size tankships and liquification plants in major petroleum-exporting countries, a world market has developed in LNG. The technology for manufacturing gas from liquid hydrocarbons is well advanced; methods for making gas from coal, though not unknown, are regarded as less satisfactory. In either case, the starting material is available in large quantities. For the period we are considering, up to 1985, SNG from coal is not likely to be important. Thus, United States gas supply, neglecting Canadian imports, will consist of production from conventional domestic sources supplemented by imported LNG and by SNG made from liquid hydrocarbons which are in turn largely imported.

In the absence of restrictions on hydrocarbon imports, the price of natural gas in the United States could be expected to rise to the price of imported LNG or of SNG manufactured from liquid feedstocks.[6] However, this is a lower-bound estimate. If in order to maintain a selected degree of self-sufficiency, the United States imposes quotas on hydrocarbon imports, then its domestic price will have to be higher in order to induce exploitation of higher-cost conventional sources within the country. Imported gas from Canada is an alternative to higher-cost gas from marginal domestic fields, so an American quota system would enhance the value of Canadian gas.

What is the value of Canadian gas likely to be? Estimates of the landed price of LNG cover a wide range, from $0.80 to $1.50 per million Btu (roughly 1 Mcf).[7] Querying the cost of SNG leads directly to the ultimate question, for which no one has an answer at present: what will be the world price of crude oil? Enough has been said, however, to permit drawing an important conclusion: Canadian exports are not likely to affect the United States price to any marked degree. We have now taken the initial step toward a new perception of the question of Canadian natural gas exports, moving from "needs" and "availabilities" to value of exports. We continue by considering the economics of supplying gas from the Canadian arctic.

B. PROSPECTIVE COSTS AND PROFITS

Information on the economics of Arctic natural gas production is not easy to obtain. In part, this is because not much is known about supply costs; no reservoirs in the far north have been brought into commercial production. On the other hand, considerable drilling experience has accumulated and a number of discoveries have been made; it should be possible to obtain some idea of production costs by making use of arctic drilling and construction cost figures, together with information about reservoir size, productivity, and gas quality where discoveries have been made. In view of both the competition that exists among oil companies and the sensitivity of the relationship between companies and resource-owning governments, it is not surprising that such information as may exist is not readily available.

A further reason for the scarcity of estimates of arctic supply cost is an understandable reluctance to publish figures which may be misused. No set of numbers describing the economics of natural gas production should be examined without a reminder of the great variability of costs in this industry. The riskiness of oil and gas production is much emphasized in accounts of the petroleum business. It is most vividly portrayed by descriptions of the great expense of drilling an exploratory, or wildcat, well and the drama of the outcome: a big discovery means wealth, or at least enhanced reputation, while a dry hole – informative though it may be – is, after all, just a hole. In reality, whether exploratory wells are productive or dry represents only one aspect of the risk to which oil men are accustomed. In translating physical exploration results into economic values, many other factors are important. For example, the depth of the productive structure strongly influences drilling expenditure and hence unit production cost; the quantity of recoverable reserves governs the extent to which possible economies of scale in production and transportation facilities can be utilized. Gas quality is of particular importance, since investment in processing plant will constitute a significant share of total investment in Arctic production facilities. Numerous additional examples could be cited of physical factors which not only affect production costs, but which can cause them to vary many-fold.[8] This situation has two significant implications:

(1) any forecast of production cost, and hence profitability, for an arctic gas field must specify a broad range of possible values, and
(2) there may be considerable variation in production costs and profitability among individual gas reservoirs within the region.

It is impossible, in spite of this warning, to understand the pressure to develop Arctic natural gas reserves without considering the economic prospects, because the *raison d'etre* of private investment is anticipated profitability.

This author has calculated production costs for arctic natural gas using various forecasts of development investment, operating outlays, and corresponding output that have appeared in published sources. Invariably, the resulting unit production costs are low relative to forecast wellhead prices.[9] An example will indicate the expectations which are held.

Figures prepared by J. R. Chilton, E. D. Bietz, and G. A. S. Bulmer of J. C. Sproule and Associates, a Calgary consulting firm, have yielded estimates of production costs for a hypothetical field in the Sverdrup Basin in the Arctic Islands, but they afford an initial insight into arctic production costs.[10] The estimates of these authors imply development cost (the component of cost which covers investment outlays) of 5.2 cents per Mcf (this cost covers a 15 per cent return on corporate capital). Implied operating cost is 4.5 cents per Mcf. The sum of 9.7 cents is to be compared with possible wellhead prices in the range of 25 to 50 cents, from which must be deducted royalty payments on the order of 2.2 and 4.3 cents, respectively. The wellhead prices are rough estimates, derived by subtracting pipeline gathering and transit costs of 75 cents per Mcf from delivered prices in the south of $1.00 to $1.25.[11]

It is quite likely that the investment figures of Chilton, Bietz, and Bulmer underestimate the outlays that will be required in the Mackenzie Delta. This seems particularly so with respect to the cost of gas processing plants. It would, however, require a 3-to 4-fold increase in total investment outlays – drilling, well completion, processing facilities – to eliminate profits with a 25-cent wellhead price, and, correspondingly, a nearly 8-fold increase at a 50-cent wellhead price. One must become accustomed to order-of-magnitude variations in petroleum production costs, but these latter outcomes would contradict widely-held expectations about gas production costs in the North.

The resource rents and resulting after-tax profits foreseen for the arctic indicate the likelihood of very attractive rewards to private companies investing in the North. The *caveat* expressed at the beginning of this section should be recalled: even if the profitability of arctic gas production on the average is very high, some ventures will be marginal or unsuccessful. This does not, however, alter the impression that exploitation of arctic natural gas holds the prospect of very high financial rewards, particularly for the large petroleum companies.

C. CANADIAN EXPORT POLICY

We have thus far ascribed economic incentives to the parties who are prominent in plans to market arctic natural gas. For prospective consumers – in particular, those represented by the United States government – the value of the gas depends on the availability of alternative sources of supply, although security is a complicating factor. The companies that will be directly involved in producing and transporting the gas respond to the profit potential of arctic operations. The remaining participant, the owner of the resources, is the Canadian government. Following the principle of symmetry, we examine the hypothesis that economic goals also underlie its policies.

At first glance, the position of the Canadian government appears more complicated than that of either of the other two parties. We noted at the outset that various groups in Canada advocate particular goals for resource policy. For example, those holding a "manifest destiny" view of regional economic growth wish to get on with resource development, while dedicated environmentalists see virtue in no development at all. Federal policy can seldom, if ever, satisfy the purist – of necessity diverse aims must be accommodated. Nevertheless, it is reasonable to hypothesize that policies are informed by articulated goals, or at least identifiable attitudes of mind, and it is in this sense that we consider the current and past record for indications that economic goals have been prominent in the evolution of federal policy toward the development of natural gas resources. We shall see that there is little support for this hypothesis.

A word of caution is necessary. We have emphasized that the current proposals for arctic resource development originate against a background of change and transition in world energy markets. Moreover, the circumstances in the arctic are novel: the federal government has full jurisdiction, without the usual division of responsibility with provincial governments. It is in circumstances of change and novelty that study of past actions offers the least basis for prediction of future actions. We will look at the genesis of the federal government's current policy, postponing until later a summary of the new circumstances with which it must now cope.

One cannot look to one ministry or agency as the source of federal policy, nor, as Pearse and Scott have emphasized in Chapter 1, is a single comprehensive policy to be found. This fragmentation complicates our investigation; while its implications for wise management of resources are profound, that subject will not detain us here.

The arrangements by which private companies produce and market the natural gas – that is, the terms of exploration permits and leases – are determined by statute and are a key aspect of federal authority; they are described in detail elsewhere in this volume.[12] Federal income tax regulations are important; they single out the resource industries to be the recipients of special tax provisions relating to the write-off of investment and the depletion of natural resource stocks. Another relevant instrument of government policy is Panarctic Oils, Ltd., in which the federal government holds a major (45 per cent) interest. Its mandate is rooted in the desire to develop Arctic resources, with the important qualification that Canadians should particpate in this development. One might locate other instances of federal authority or activity, for example, subsidization of studies of transportation in the arctic. We pass over these to the agency which has wide statutory authority over the distribution and sale of natural gas, the National Energy Board. The NEB was established following a comprehensive study of the goals of federal energy policy, and it occupies a strategic position by virtue of both the specific instruments by which it regulates natural gas production and its broad mandate to study and coordinate all federal energy policy.

The NEB's mandate originated with the Royal Commission on Canada's Economic Prospects in 1957. In consequence of the separate study of energy prospects, the Commission's final report contained the following recommendation:[13]

> In order that a sound and comprehensive policy may be worked out with regard to development, exports, imports and consumption of all forms of energy in Canada, we propose that a national energy authority be established which would be responsible for:
> (a) advising the Federal Government and, upon request, any provincial government in all matters connected with the long-term requirements for energy in its various forms and

in different parts of Canada; methods of promoting the best uses of energy sources from a long-term point of view; export policy, including such questions as the further refining of oil and gas in Canada and the disposal of by-products; coal subsidies, etc.

(b) approving, or recommending for approval, all contracts or proposals respecting the export of oil, gas and electric power by pipeline or transmission wire, including, where necessary or desirable, the holding of public hearings in connection therewith.

In response to the recommendations of this commission, another was created, the Royal Commission on Energy, more commonly known as the Borden Commission. It was charged with investigating means for insuring "that present and future Canadian requirements for energy are taken fully and systematically into account in granting licences for the export of energy or sources of energy."[14] The recommendations of this second commission, as enacted into law, established the National Energy Board and outlined its regulatory and advisory functions. A succint summary of these functions is given by the introduction to the Board's most recent annual report:[15]

It (Parliament) established the Board as a regulatory authority responsible for adjudicating upon applications for licences to export and to import natural gas, applications to construct and operate interprovincial and international pipe lines for the transportation of petroleum and its products and of natural gas, and application for licences to export electrical power and energy and for certificates to construct and operate international power lines. It also assigned to the Board the responsibility for regulating the rates, tolls and tariffs of oil and gas pipe lines under the Board's jurisdiction.

Parliament also gave the Board an advisory role. In fulfilling this responsibility the Board is required to keep under review all matters relating to energy within the jurisdiction of the Parliament of Canada and to recommend to the Minister any measures that the Board considers necessary or advisable to undertake in the public interest to ensure proper use and development of energy and its sources.

The Board's domain has since been broadened but we will be concerned with policies arising in the implementation of the original statute.

With regard to natural gas, the NEB's attention is focused upon the granting of export permits. In this it accurately reflects the thinking of the original Royal Commission: Canadian energy resources should be developed with regard to Canadian needs, adn this goal may be frustrated by excessive exportation. The Board is guided by the National Energy Board Act, which declares:[16]

> Upon an application for a licence, the Board shall have regard to all considerations that appear to it to be relevant and, without limiting the generality of the foregoing, the Board shall satisfy itself that
>
> (a) the quantity of gas or power to be exported does not exceed the surplus remaining after due allowance has been made for the reasonably foreseeable requirements for use in Canada having regard, in the case of an application to export gas, to the trends in the discovery of gas in Canada;
> . . .

The Board thus faces the task of implementing the formula: available reserves – requirements = exportable surplus.

The NEB's procedures are described in its reports of decisions on particular export applications. Table 2 illustrates the accounting that is made of requirements and available reserves. We will not dwell upon the difficulties involved in estimating natural gas needs: consumption forecasts must take account of substitution among fuels in response to relative prices, which in turn reflect changing supply costs and changing uses. To arrive at its figures the Board first projects the sales of gas over a 30-year period. Requirements are then computed for Alberta as 30 times the quantity needed to meet consumption in the current year; for the rest of the country the requirements are computed as 25 times projected consumption four years hence. To these figures are added several other items, the most important being existing and requested exports. This type of calculation is repeated for future years.

On the stock side the NEB estimates the quantity of established reserves which are currently available. These consist of all proved reserves and a fraction, usually around 50 per cent, of probable reserves. Proved reserves actually comprise the overwhelming volume of available reserves; in 1969, for example, the Canadian Petroleum Association's estimate of proved reserves in Canada amounted to slightly over 95 per cent of the NEB's corresponding figure for available established reserves. Available reserves in the future are projected by assuming that additions take place at the average trend rate observed in data from the past ten years. The NEB reserves figures are appropriately likened to warehouse inventory, as will be evident when we examine resource stock measures in greater detail.

If the difference between "available reserves" and "total requirements" is positive, an exportable surplus exists. The Board – after scrutinizing such details as the financial responsibility of the companies involved and the sales price (to see that it is at least as high as the geographically comparable domestic price) – may then grant an export licence. In 1971 the NEB rejected applications that would have provided for the export of 2.66 t.c.f. of natural gas, judging that Canada did not have this quantity of exportable surplus. It had denied or modified applications in the past, but this decision was precedent-making in its clearcut insistence that stocks had to meet its accounting criterion before exports would be allowed.

In putting into practice its authority to regulate natural gas exports, the NEB's concentration on needs and availability is reminiscent of the NPC report.[17] Such a policy framework contrasts with an economic one: attention is directed to things rather than to values. This is an ancient source of confusion. Karl Marx, whose own value theory had its shortcomings, nonetheless clearly saw the pitfall of focusing on physical quantities, which he memorably characterized as the "fetish of commodities." Marx recalled the mercantilist's attachment to gold and silver and "the physiocratic illusion that land-rents are a growth of the soil."[18] Some useful information can be gained from stock measures of natural resource commodities, but the hazard of looking only at such measures in the case of natural gas is that there may be more important things to watch – in particular, the creation and division of the incomes which derive from economic values.

In the concluding section of this paper we will sketch some of the economic implications of large-scale resource exports. Dollar scorekeeping draws attention to important issues that are not revealed when the tally is in standard cubic feet. First, however, we will digress long enough to see what the NEB learns from its physical measures of resource availability.

D. THE QUANTITY APPROACH TO RESOURCE MANAGEMENT

The NEB, in striving to ensure that natural gas exports do not reduce available supplies below estimated Canadian needs,

Table 2
Forecast of Future Relationship Between Requirements and Supply
(Tcf at 1,000 Btu/cf)

Years commencing 1 January	1970	1975	1980	1985	1990
Requirements					
Alberta	7.5	9.7	12.0	14.8	17.8
Canada, Except Alberta	25.9	32.7	39.9	48.3	56.8
Processing Shrinkage	2.2	2.4	2.6	2.9	3.2
Existing Exports	12.0	8.3	4.9	2.2	0.6
Requested Exports	8.9	7.6	5.5	3.2	0.8
Cumulative Canadian Consumption	-	6.2	14.2	24.1	36.0
Cumulative Exports	-	5.1	10.6	15.6	19.6
Cumulative Processing Shrinkage	-	0.2	0.4	0.6	0.7
Total Requirements	56.5	72.2	90.1	111.7	135.5
Supply					
Established Reserves at 31 Dec. 1969	57.4				
Less Deferred Reserves	(1.7)				
Less ½ Beyond Economic Reach	(1.8)				
Plus Imports	0.1				
Available at 31 December 1969	54.0	54.0	54.0	54.0	54.0
Plus Historical Growth at 3.5 Tcf/yr	-	17.5	35.0	52.5	70.0
Sub-total	54.0	71.5	89.0	106.5	124.0
Plus Additional Growth Required from 1970	-	0.7	1.1	5.2	11.5
Total	54.0	72.2	90.1	111.7	135.5
Surplus (Deficiency)	(2.5)				
Average Required Growth from 1970, Tcf/yr		3.6	3.6	3.9	4.1

Source: National Energy Board. *Report to the Governor in Council . . .* , August 1970, p. 4-40.

resembles a prudent granary manager. Foreign buyers are welcome at the granary when bumper harvests have filled the storage bins, but after lean years of drought, the granary stock must be husbanded to feed the home population. Whereas fluctuations in the grain harvest usually are attributable to variations in climate, it might be supposed that changing fortunes in prospecting periodically generate surpluses of natural gas reserves. At such times the NEB might approve export contracts.

Time-series data of the sort used by the NEB reveal that periodic fluctuations in stock level, as required by the granary analogy, are not characteristic of the natural gas industry. Before examining the behaviour of these data, it is necessary to specify precisely the nature of the stock being measured. It has already been noted that the NEB's established reserves comprise proved reserves and probable reserves, with the latter accounting for only a very minor fraction of the total. These two types of reserves are defined as follows:

> *Proved Reserves* consist of the gas recoverable from the area of a pool completely delineated by drilled wells. A portion of the reserves may be under undrilled spacing units which are so close and so related to drilling spacing units that there is every reasonable probability they will be produced when the spacing units are drilled or will be recovered from existing wells.

> *Probable Reserves* consist of the gas estimated to be recoverable from a pool beyond the proved limits of the pool. The probable pool limits are based on normal geological expectation.[19]

Proved reserves represent warehouse inventory because the drilling of wells simultaneously defines, or creates, proved reserves and established producing capacity. Accordingly, proved reserves meet two tests: (1) their existence is known with virtual certainty, and (2) they are producible with existing technlogy under existing economic conditions. Probable reserves do not meet the first of these tests, although their existence is regarded as highly likely; they are presumed to be potentially producible under current economic conditions, but the requisite development wells have not been drilled.

Table 3 provides information on proved reserves of natural gas for the three provinces which have accounted for nearly all Canadian production. Stock levels are shown as the ratio of proved reserves to production (sometimes misleadingly called "the life index"). Measured by this ratio, stock levels in the past have been relatively much higher than at present. Most striking is Alberta, by far the largest producer. There the ratio of proved reserves to production was almost 80 in 1957 and 1958, but it had declined to under 30 by 1970. This secular decline completely over-shadowed year-to-year variation. In Saskatchewan the pattern of decline was repeated, though it was not so pronounced. British Columbia differed in that the ratio rose prior to 1964, declined subsequently, and has displayed some year-to-year variation.

The notion that an excess, or surplus, of reserves is a periodic result of changing fortunes in exploration is inconsistent with the continuous decline in the ratio of proved reserves to production which is observed. A more plausible explanation for the existence of surplus stocks lies in the joint occurrence of crude oil and natural gas. Natural gas is nearly always produced with crude oil (though the converse is not true). Exploration activity may yield either crude oil or nonassociated gas. As the western provinces were searched for crude oil, an inventory of nonassociated gas fields was accumulated.

Before natural gas demand was developed to its present degree, gas produced jointly with crude oil was plentiful. Nonassociated gas fields were either "capped" (that is, held without installation of producing capacity) or, if developed, were drilled on a very wide spacing pattern. With the growth of demand for gas and the laying of additional trunk pipelines, the undeveloped fields were brought into production and greater capacity was obtained from fields already producing by means of more intensive drilling. The result was a period during which output could be, and was, increased at a faster rate than new reserves were discovered.

As indicated in Table 3, in 1970 the fraction of gas reserves consisting of nonassociated gas was over 97 per cent for British Columbia, nearly 90 per cent for Alberta, and nearly 70 per cent for Saskatchewan. For Alberta this represented a small but steady increase over the decade shown, while for Saskatchewan the increase was more pronounced. In British Columbia, where little crude oil has been found, petroleum exploration must be largely sustained by the expectation of gas discoveries, a situation which prevails in much of Alberta as well.[20] To the extent that gas reserves are held independently of crude oil, excess reserves (or surplus inventory) are an anomaly. With gas, as with other commodities, creating and holding large inventories is costly. It is

TABLE 3
Reserves of Natural Gas: Ratio of Proved Reserves to Production, Type, and Value Alberta, British Columbia, Saskatchewan

YEAR	58	60	62	64	66[4]	68	70
ALBERTA							
Ratio of Proved Reserves to Production[1]	78.7	64.5	40.0	38.9	39.40	35.6	27.9
Nonassociated Reserves as Percent of Total Reserves[2]			83.1	86.0	86.8	86.5	87.3
Average Field Sales Value, Cents Per MCF[3]	7.8	7.8	10.8	12.5	15.2	15.6	16.4
BRITISH COLUMBIA							
Ratio of Proved Reserves to Production[1]	27.0	37.7	43.3	50.0	41.2	31.2	38.3
Nonassociated Reserves as Percent of Total Reserves[2]			98.6	98.4	98.0	98.0	97.5
Average Field Sales Value, Cents Per MCF[3]	5.5	8.7	9.0	8.8	9.8	10.2	11.8
SASKATCHEWAN							
Ratio of Proved Reserves to Production[1]	26.3	26.5	17.8	18.1	16.2	15.7	19.7
Nonassociated Reserves as Percent of Total Reserves[2]			50.7	53.9	52.7	53.4	69.3
Average Field Sales Value, Cents Per MCF[3]	3.8	6.4	4.9	6.1	11.1	14.0	15.2

Sources:
(1) Proved reserve and production figures from *Reserves of Crude Oil, Natural Gas Liquids and Natural Gas in the United States and Canada*; published by American Gas Association, American Petroleum Institute, and Canadian Petroleum Association; various years.
(2) *Ibid*
(3) Values of sales from Canadian Petroleum Association, *Statistical Yearbook*, 1970
(4) Production and reserves figures reported as recoverable reserves through 1965; beginning with 1966 figures are marketable reserves.

not surprising, therefore, that the relative quantity of proved reserves should have shrunk in recent years. The economically optimal stock of proved reserves depends on such factors as wellhead price and production cost. It is possible that at the average reserves-to-production ratio that now prevails reserves are still uneconomically large; if so, the average ratio could be expected to continue to fall in the absence of government intervention.[21]

When the NEB refused to grant export licences in 1971 because its figures showed that no exportable surplus existed, its behavior was consistent with that of the prudent granary manager. Warehouse inventory did not come up to estimated needs. However, the NEB's criteria for selection of the stock levels deemed necessary are not explicit; no analysis has been offered (to our knowledge) of the economic cost of holding inventory or the cushion needed to safeguard against the risk of stock shortage caused by a run of unfavorable exploration results. One suspects that the Board's willingness to adopt an arbitrary rule-of-thumb for dealing with conventional inventory issues results not so much from insufficient attention as from a conviction that such matters are not its central concern.

The supply of natural gas, unlike the grain in our example, is ultimately limited to the amount now deposited in the earth's crust. If exhaustion of resource stocks is the main worry of the NEB, then the granary analogy is misplaced. In these circumstances, however, the NEB should not focus its attention on statistics which measure available stock; rather it should look to measures of the remaining stock of natural gas in place in Canada.

The term "resource base" is used to describe "the total amount (of the resource) present in the earth's crust within a given geographic area."[22] This concept of scarcity is advocated on the ground that, over time, experience has shown the resource base to be the constraint imposed by nature. The reserves *now* known and considered economic represent only a fraction of the reserves to be drawn upon following further exploration activity, raw material price increases, or improvements in extraction technology.

Estimates of the resource base are, of course, subject to a larger margin of error than the other measures considered so far. One set of estimates, prepared by the Canadian Petroleum Association (CPA), is shown in Table 4. The figures show the "total potential reserves of natural gas that could be expected to ultimately be recovered in Canada, using present day technology and conventional methods of oil and gas production."[23] These estimates understate the resource base as just defined, since the CPA does not allow for the possibility that new technology will increase the fraction recoverable as well as permitting the exploitation of accumulations not now accessible. A more recent set of estimates of potential Canadian reserves, prepared by a private company, is also shown in Table 4. Not enough is revealed about the derivation of these estimates to indicate whether the substantially higher total is attributable to a more optimistic bias or to the accumulation of new information, although the latter is probably important in view of the successes being achieved in arctic exploration. Table 4 contains some additional data on production, proved reserves, and ultimate reserves for purposes of comparison. The figure for ultimate recoverable raw reserves (97.82 t.c.f.) provides an indication of the quantity of natural gas in fields already discovered; this amount represents 13.5 per cent of the CPA estimate of potential reserves and 9.3 per cent of the higher Dome estimate.

Although the CPA estimates were prepared at the instigation of the NEB, there is no evidence that they have influenced NEB policy. Moreover, it is difficult to see how this type of stock measure could be used in a system of physical accounting. The NEB, as already noted, applies rather mechanical rules in making use of its "established reserves" inventory. Rules-of-thumb defining the rate of use of an estimated resource base would be more difficult to justify and apply. The potential reserves are defined without regard to cost, and to prescribe a rate of use ignoring economic considerations would create severe strains.

The NEB's goal of maintaining some degree of comparative abundance of natural gas in Canada is thus not as uncomplicated as it might first appear. There are, as we have seen, several measures of resource stocks, roughly graduated by degree of certainty. Which of these is appropriate as an index of resource availability depends on whether concern is focused upon security of supply in the foreseeable future or upon the prospect of eventual physical scarcity; a measure that pertains to the former may misrepresent the latter. However, rather than considering the most suitable way of applying a commodity-oriented policy to arctic natural gas development, we will return to the main line of our discussion. Taking account of the other interested parties in the arctic, what are the implications of the Canadian government's continued pursuit of the goals it has heretofore sought? In particular, what other considerations may be obscured?

E. IDENTIFYING THE PUBLIC INTEREST

We have now considered each of the three principal players with a stake in arctic natural gas. The United States has been widely depicted as desperately searching for new energy supplies to meet its needs, a description that conveys an overtone of "without regard for cost." However, over the longer period which is appropriate in a discussion of arctic resource development, the United States has many supply alternatives. A period of adjustment among energy sources will be rendered painful by a marked increase in the average cost of energy, but during this transition the relative cost of various sources will guide the choice of fuel mix, in spite of some significant deviations resulting from government intervention in the name of security.

The position of the companies with interests in arctic resource development is straightforward. Private corporations seek basically to maximize the return on invested capital, and there is nothing in the current circumstances to suggest that this incentive has been displaced by another.

The third player, the Canadian government, because of its ownership of arctic resources and its authority over interprovincial and international trade, is in a position to influence the terms under which development of natural gas resources takes place and the rate at which development proceeds. With regard to the terms of alienation, the federal government has considerable, though by no means unlimited, power. Were it to seek to progressively increase its revenues from natural gas production (by any combination of higher taxes, lease bonuses, or royalties), there is a point at which the companies would choose to invest elsewhere; this is because they face a limit on the price Americans will pay, given their alternative supply sources.[24] The record, however, suggests that the Canadian government's primary interest has not been directed toward its potential share of resource income; instead, the desire to foster development of the arctic has resulted in concession terms very favorable to resource exploitation (See Chapter 5). With regard to the rate of development, the government can impose restraint. Precedent is provided by the NEB, which has controlled exports of natural gas (and hence the rate of exploitation) according to standards it has set up to define Canadian needs and availabilities. This approach, as we have seen, implements the goal of maintaining Canada's position of relative resource abundance.

We remarked earlier that because the circumstances of arctic resource development are novel, past behavior may be a treach-

TABLE 4
Potential Reserves of Natural Gas in Canada*

Canadian Petroleum Association[1]		Dome Petroleum Ltd.[2]	
Area	Quantity (trillion cubic feet)	Area	Quantity (trillion cubic feet)
Western Canada Sedimentary Basin	270	Western Canada	300
Arctic Islands, Coastal Plain, Foxe Basin	261	Arctic Islands	300
		Mackenzie River	
East Coast Offshore	150	Valley	210
Other	44	Other	240
Total	725	Total	1050

Reference Data: Canadian Production, Remaining Marketable Reserves, and Ultimate Reserves[3] (trillion cubic feet)

Net production, 1969	*1.6*
Cumulative net production to Jan. 1, 1970	*11.6*
Proved remaining reserves, Jan. 1, 1970	*51.9*
Proved plus probable remaining reserves, Jan. 1, 1970	*57.5*
*Ultimate raw reserves (proved plus probable), Jan. 1, 1970***	*97.8*
*Ultimate marketable reserves (proved plus probable), Jan. 1, 1970***	*76.8*

 * Potential gas reserves refer to already produced, existing proved and probable, and yet-to-be-discovered reserves recoverable using present-day technology and conventional production methods.
** Ultimate recoverable reserves are defined as "the total quantity . . . estimated to be ultimately producible from an oil or gas field as determined by an analysis of current geological and engineering data. This includes any quantities already produced." Marketable gas differs from raw gas in that "certain hydrocarbon and non-hydrocarbon compounds have been removed or partially removed by processing."

Sources:
1. Canadian Petroleum Association Geological Reserves Committee, *Potential Reserves of Oil, Natural Gas and Associated Sulphur in Canada* (Calgary, 1969), p. 4-6.
2. J. C. McCaslin, "What they've found in the Arctic," *Oil and Gas Journal* (October 23), 1972, p. 69-78.
3. American Gas Association (jointly with American Petroleum Institute and Canadian Petroleum Association), *Reserves of Crude Oil, Natural Gas Liquids, and Natural Gas in the United States and Canada* . . . vol. 26 (May 1972), pp. 229, 234, 241, 242.

erous guide to future government policy. In the arctic the federal government does not share authority with the provincial government as it is in the habit of doing elsewhere. Furthermore, rapidly changing conditions in various energy resource markets have revealed the obsolescence of many existing attitudes and arrangements, a notable example being the National Oil Policy. It is possible, therefore, that current policy reviews in Ottawa may lead to new objectives and new procedures.

In other areas of responsibility the federal government is accustomed to applying economics when confronted with a choice among policy alternatives. It does so not only in formulating stabilization and trade policy, but also in the regulation of transportation and the budgeting of specific projects. In the latter areas the accepted goal is that resources be allocated to projects that make a positive contribution to national welfare, hopefully selecting where alternatives exist those projects which make the largest contribution. Contribution to welfare is gauged by applying some form of economic accounting, that is, by systematic comparison of benefits with costs. It is instructive to consider arctic resource development in these terms.

In the exploitation of natural gas resources there are social costs as well as conventional production costs. Examples which have been cited (and which receive more detailed examination in other papers in this volume) are the relocation of workers for what may be temporary employment, the dislocation of native people, and damage to the environment. Identification of social costs becomes particularly difficult where a project may be sufficiently large to generate perceptible macroeconomic effects. An example is the claim that development and sale of arctic petroleum resources will have a significant effect on the balance of payments with adverse long-run consequences for Canada's industrial structure. There will be disagreement when evaluating these costs and devising ways by which they can be met or countered, but the general principle – that such costs exist and must be compared with benefits – is not controversial.

A great deal of confusion surrounds the identification of benefits. Income accruing to the federal government as a result of natural gas production is a social benefit; it can be used to ameliorate adverse side-effects of arctic development. At the other extreme, it is difficult to see much benefit for Canada in profits (that is, above-normal returns on invested capital) accruing to foreign corporations. Not quite so obvious is the social benefit to be attached to profits (again, defined as above-normal returns on invested capital) accruing to Canadian corporations. However, the proposition that one dollar of such corporate profits is of equal social benefit to one dollar accruing to the federal government is not one that would command widespread support as a guideline for the division of arctic resource rents.

The abbreviated sketch in the preceding paragraphs illustrates some of the questions that are raised by an economic approach to social accounting, which we referred to earlier as dollar score-keeping. The comprehensiveness of an economic approach to defining Canadian interests in the arctic is in itself a departure from the federal government's *modus operandi,* which has up to now been the pursuit of several diverse goals in juxtaposition. More striking than this, however, is the emergence of issues which have so far been ignored or relegated to the background. The final judgment on whether development of arctic natural gas resources is a good thing for Canada will depend on which way the Canadian government is looking when it happens.

4.

FORECASTING NORTH AMERICAN
ENERGY DEMAND:
ISSUES AND PROBLEMS

Ernst R. Berndt

Much of the research into the economics of energy has focussed on the problems of supply. Researchers have typically addressed themselves to questions such as, what are the costs and probable successes of exploration? Once a discovery has been made, what are the relations among cost, level of output, and grade of output? What are the environmental costs associated with the generation, transmission and distribution of energy? At what price will new technologies become economically feasible?

Comparatively little research, however, has been directed to analysing the demand for energy. Research on demand for energy might consider the following issues: Why is North American energy demand increasing so rapidly? How responsive is the quantity of energy demanded to a change in its price? How would increases in the relative prices of heating oil and natural gas affect the welfare of people in low, medium and high income brackets? Do federal investment tax incentives designed to curb unemployment affect industrial demand for energy?

In this short paper we focus attention on factors that affect the demand for energy. We briefly review what is known about the demand for energy, and then we evaluate several recently published projections of future energy demands for Canada and the United States. Our attention here is restricted to the demand for energy; it must be emphasized that any analysis of the North American energy market depends equally on considerations of supply – which are dealt with in other papers in this volume.

A. ANALYSING THE GROWTH OF ENERGY DEMAND

Most of the early research on the demand for energy was confined to the United States. One of the first major studies was produced in 1960 by Resources for the Future, a private research group originally funded by the Ford Foundation. This study was an extensive analysis of the supply and demand for energy in the United States from 1850 to 1955, and it included projections through 1975; it marked the first time that researchers had assimilated, published, and analysed historical statistics on energy demand. It was followed, in 1971, by an analysis of world supply and demand for energy.[2]

One of the most interesting findings of the 1960 and 1971 studies was that of a systematic relationship between total energy consumption and the Gross National Product – summarized by the "energy-GNP ratio." From 1920 to 1960 the energy-GNP ratio fell at a fairly constant rate in both Canada and in the United States. This finding has significant implications, for if we can depend on a stable trend in the energy-GNP ratio, then it is fairly easy to forecast energy demand by simply multiplying the projected value of GNP in some future year by the energy-GNP ratio.

Recent events suggest that this method of projecting energy demand may be unreliable, however. Between 1966 and 1970, the energy-GNP ratio in the United States reversed its long-term historical decline and began to rise; then in 1971 and 1972 its value fell back to the level of the early 1960's. Since energy projections were based on the assumption of a relatively stable or "time-trended" ratio, this instability of the ratio cast considerable doubt on the reliability of the projections.

To an economist, it is not at all obvious why one would want to focus on the energy-GNP ratio rather than, say an energy-capital stock ratio or an energy-labour ratio. Indeed, the results of a recent empirical study suggest that energy demand is more closely correlated with capital than with GNP or labour.[3] Furthermore, even if an economist were persuaded to analyse one of these ratios, the first point he would probably make is that the ratio would be expected to vary over time in response to changes in tastes, technology, and relative prices.[4] For example, if he were to examine postwar trends in the energy-GNP ratio he might well conclude that its increase simply indicates that users of energy respond to changes in its price, for over this period the relative price of energy has fallen and energy has become an increasingly good bargain.

If energy demand responds to changes in its price, projecting future energy demand on the basis of the energy-GNP ratio is highly misleading; this method would be appropriate only if either relative prices were to remain constant in the future, or if "price doesn't matter," i.e. if the quantity of energy demanded is not responsive to changes in relative prices.[5] Let us consider these two possibilities briefly.

Most observers believe that the relative price of energy will increase in the next decades. Indeed, it is commonly believed by other contributors to this volume, among others, that recent shortages of natural gas in the United States are due to the Federal Power Commission's ceiling on the price of natural gas— a ceiling well below the price that would hold if the market forces of supply and demand operated freely. It appears likely that this ceiling will be lifted soon; if it is, it is likely that the price of natural gas will double or triple. And, since electric utilities are large consumers of natural gas, natural gas price increases will most likely result in price increases for electricity. Furthermore, recent agreements between the oil exporting countries of the Middle East and producing companies have already led to price increases for crude oil, and it is expected that this escalation will continue into the 1980's. Finally, environmental restrictions are expected to exert substantial upward pressure on the prices of energy. In short, it appears that the relative price of energy will rise in the coming decades. This implies that projecting energy demand by the method of the energy-GNP ratio is appropriate only if "price doesn't matter."

The available empirical evidence on the importance of prices suggests that the demand for energy is in fact quite responsive to price. H. Houthakker and P. Verleger have recently released results of a study on the demand for gasoline in the United States indicating that if the relative price of gasoline increases by

1 per cent, the quantity demanded declines by about .4 of 1 per cent in the short run and in the longer run by about .75 of 1 per cent.[6] Evidence concerning the price responsiveness of electricity demand, published by Chapman, Mount, and Tyrell, suggests that if the relative price of electricity were increased by 1 per cent, the quantity of electricity demanded would decrease in the longer run by 1.1 per cent in the residential sector, by 1.3 per cent in the commercial sector, and in the industrial sector by 1.5 per cent.[7] Similar results on the price responsiveness of electricity demand have been obtained by other researchers.[8] It has also been found that the demand for natural gas is price responsive.[9] In addition, Chapman, Mount, and Tyrell found that most published forecasts of electricity demand in the United States were based on population growth rates that were considerably greater than the recent historical pattern. Taking account of probable increases in the relative price of electricity and likely growth rates of population, Chapman, Mount, and Tyrell conclude that the published projections have substantially overestimated future demand for electricity.

An important finding in each of the above studies is that the quantity of energy demanded in the residential sector increases with family income, but that the *proportion* of the family's budget spent on energy is smaller the higher its income. This implies that when increased taxes, environmental restrictions, removal of price ceilings, or other factors bring about an increase in the price of energy, the burden of this price increase will be borne more heavily by the poor. It is likely, therefore, that unless the government takes compensating policy actions, the poor will suffer disproportionately from probable increases in the relative price of energy.

The studies cited above all conclude that energy demand is quite responsive to changes in its price. This is a comforting finding, for it suggests that governmental policies which affect the price of energy can be effective instruments in altering the nation's demand for energy. For example, a tax on energy such as that recently proposed (but not implemented) in Ontario, would increase the price of energy to consumers and thereby reduce the quantity demanded.

Energy taxes are only one example of governmental policies that might significantly alter the composition and level of future energy demand. An important finding of a recent study on industrial energy demand is that energy and labour are substitute inputs in production processes, while energy and capital are complementary.[10] This means that if the price of energy were to rise more rapidly than that of labour, industrial users would tend, in the long run, to employ more labour and less energy and capital for a given level of output. In this respect rapid price increases for energy may be good news to policy makers concerned with Canada's unemployment problems, for if energy becomes increasingly more expensive, firms will tend to substitute labour for energy and thereby increase the aggregate demand for labour.

The outlook for capital, however, is less favourable. In recent years the federal governments of Canada and the United States have frequently offered tax incentives to investors as fiscal instruments to stimulate the growth of output and the reduction of unemployment. Since capital and energy are complementary inputs, the effect of corporate investment tax incentives is to increase the demand not only for capital but for energy also. In the future, however, federal governments may be less inclined to use this kind of fiscal incentive, because such incentives would result in greater demand for increasingly scarce energy. Economic policy critics such as Eric Kierans have also argued, incidentally, that these investment incentives actually add to Canada's labour unemployment problems by making the economy more capital intensive.[11]

We can expect then that future governments will try to develop new policy instruments to stimulate economic activity, which do not add such pressures to energy demand. One possible device is (instead of an investment tax incentive) a labour tax credit – a subsidy given to employers if they hire new labour. Such a policy would decrease the price of labour to employers. Since labour and energy are substitutable goods, a labour tax credit would stimulate employers to use more labour and less energy and capital.

We have argued above that the relative price of energy is likely to rise, that the demand for energy is price responsive, and that future energy demands can be altered significantly by governmental tax policies that affect the relative prices of energy, labour, and capital. Since future energy demand will be affected by prices and tax policies, forecasts of future energy demand based solely on energy-GNP ratios must be considered unreliable.

B. ARE FORECASTS RELIABLE?

In this section we briefly summarize methods of energy projection employed in three recent studies – two Canadian and one from the United States: (i) a 1969 National Energy Board(NEB) forecast of energy supply and demand in Canada for 1975 and 1990,[12] (ii) a 1973 Energy, Mines, and Resources (EMR) projection through 2000,[13] (iii) an "industry forecast" of United States energy demand through 1985, published in 1971 by the National Petroleum Council (NPC).[14] The NEB and NPC studies take no account of possible changes in prices. For this reason the reliability of their projections is questionable. In contrast, the EMR study considers the effect of rising energy prices on both energy demand and supply.

Forecasts of energy demand are typically prepared for the end-use sectors – transportation, residential, commercial and industrial – and then aggregated to arrive at a total secondary energy demand forecast. In order to arrive at a forecast of total primary energy demand, the conversion losses incurred in supplying secondary energy demand must also be estimated.

We begin with industrial energy demand. The NPC study uses the energy-GNP ratio method, and assumes that industrial production will increase at an annual rate of 4.4 per cent. The NEB projection is based on extrapolation from the observed trend in industrial energy demand over past years. Neither study introduces price as a variable affecting industrial energy demand. The EMR projection is based on separate forecasts for each of five industry groups: the pulp and paper industry, the chemical industries, the iron and steel industry, the metal smelting and refining industries, and all other industries. The authors of this report frequently mention the important role of price, and allow for some moderation of demand due to rising energy prices. Precisely how price was incorporated into the projections is not explained, however. Although several variations of the standard forecast are considered, the EMR study does not discuss how future energy demands would be affected if energy prices increase slowly, moderately, or very rapidly. Furthermore, like the NPC and NEB studies, the EMR projection ignores the role of investment tax incentives on future energy demand. We conclude that the EMR forecast of industrial energy demand is clearly preferable, but that considerably more work must be done to make these projections sufficiently reliable.

The NPC forecast of residential and commercial energy demand is simply based on a time-trend extrapolation. The NEB projection is more detailed, for it is based on the projected number and composition of households and household dwellings (which leaves the projections quite sensitive to future estimates of

EMR Standard Forecast of Canada's
Primary Energy Consumption

(10^{15} BTU)	1970	1980	1990	2000
Petroleum	3.1	4.8-5.4	6.7-10.2	9.4-14.6
Natural Gas	1.2	3.1-2.5	6.0-2.5	8.2-3.0
Coal	0.7	1.1	1.6	1.9
Hydro Electricity	1.5	2.3	3.1	3.4
Nuclear and Other	—	0.5	2.0	5.1
Total	6.5	11.9	19.4	28.0

(Natural Units)

	1970	1980	1990	2000
Petroleum (million barrels per day)	1.5	2.3-2.5	3.2-4.8	4.4-6.9
Natural Gas (trillion cubic feet)	1.2	3.1-2.5	6.0-2.5	8.2-3.0
Coal (millions of tons)	26	49	84	100
Hydro Electricity (billions of Kwh)	157	235	310	344
Nuclear (billions of Kwh)	1	45	180	502

birth rates). Neither the NPC or NEB forecast takes account of the changes in the price of energy or of personal income levels. The EMR study separately forecasts residential home-heating demand, other residential uses, and commercial demand. The home-heating demand forecast is based on the assumption that energy consumption for heating will remain constant per household, so that aggregate demand for this purpose in future years depends only on the estimated number of households. No account was taken of probable increases in the efficiency of energy use result-ing from higher future energy costs. Since the EMR study assumes that the price of electricity in real terms will not change, it projects residential non-heating energy demand on the basis of time-trends and expert judgement. Its forecast of commercial energy demand is simply an annual growth rate of 9 per cent in this decade, 6 per cent throughout the 1980's, and 4 per cent throughout the 1990's. The basis of these projections is not given in the EMR report; the study acknowledges, however, that because of the lower projected price increase for electricity, electricity is likely to account for an increasing share of total commercial energy consumption in the future.

The NPC forecast for the transportation sector is based on a simple time-trend extrapolation, while that of the NEB study is based on projected future motor vehicle registrations. In the EMR study, gasoline consumption by automobiles was forecast by estimating population, automobile-population ratios, and yearly gas consumption per automobile. Automobile-population ratios were estimated to increase through the year 2000, while gasoline consumption per vehicle was assumed to remain constant at 700 gallons per year. Price and income effects were not explicitly considered. Energy demands for rail, air, and marine transportation were estimated by time-trends and expert judgment.

The salient point emerging from this brief summary of projection methods is that the NPC and NEB studies are of questionable reliability and usefulness because they take no account of the effects on demand of changes in energy prices. While the EMR study recognises the important role of prices, it never states explicitly how energy price will effect future demand. If one must choose from among these forecasts, the EMR projection is clearly preferable. This forecast is summarized in the accompanying table. However, we concur with the authors of the EMR report who conclude that "At best the forecasting of energy demand and supply is a hazardous exercise, but a continuing program of periodical review can improve the quality of forecasts."[15]

FOOTNOTES

1. Sam H. Schurr, Bruce Metschert, et. al., *Energy in the American Economy, 1850-1975* (Baltimore: John Hopkins Press, Resources for the Future, Inc., 1960).
2. Joel Darmstadter, et. al., *Energy in the World Economy* (Baltimore: John Hopkins Press, 1971).
3. Ernst Berndt and David O. Wood, "Tax Policy and the Derived Demand for Energy Input," University of British Columbia (mimeo) October 1973.
4. To an economist, therefore, the everyday phrase of Canada's "future energy requirements" is misleading, at best. Canada's future energy demands will be determined in part by the kind of society Canadians want to build in the future. The amount of energy required in the future is still largely a matter of choice.
5. For a more detailed discussion, see Ernst R. Berndt and David

O. Wood, "An Economic Interpretation of the Energy-GNP Ratio," chapter 3 in M. Macrakis, editor, *Energy: Demand, Conservation, and Institutional Problems* (Cambridge: MIT Press, 1974).

6. H. Houthakker and P. Verleger, "The Demand for Gasoline: A Mixed Cross Sectional and Time Series Analysis," paper given at the December 1973 meetings of the American Economic Association, New York.

7. Duane Chapman, Timothy Mount, and Timothy Tyrell, "Electricity Demand Growth: Implications for Research and Development," testimony prepared for the Subcommittee on Science, Research, and Development of the Committee on Science and Astronautics, United States House of Representatives, June 16, 1972.

8. See, for example, two papers by Kent P. Anderson: "Toward Econometric Estimation of Industrial Energy Demand: An Experimental Application to the Primary Metals Industry," Rand Corporation Report R-719-NSF, December 1971, and "Residential Demand for Electricity: Econometric Estimates for California and the United States," Rand Corporation Report R-905-NSF, January 1972.

9. Pietro Balestra, *The Demand for Natural Gas in the United States – A Dynamic Approach*, Series Contribution to Economic Analysis (Amsterdam: The North-Holland Press, 1967).

10. Ernst Berndt and David O. Wood, "Technology, Prices, and the Derived Demand for Energy Input," mimeo (University of British Columbia; Dec. 1973).

11. Hon. Eric Kierans, P.C., M.P., "Contribution of the Tax System to Canada's Unemployment and Ownership Problems," Address to the Annual Meeting of the Canadian Economics Association, St. Johns, 1971, (mimeo).

12. National Energy Board, *Energy Supply and Demand in Canada and Export Demand for Canadian Energy, 1966 to 1990* (Ottawa: Information Canada, 1971).

13. The Department of Energy, Mines, and Resources, *An Energy Policy for Canada – Phase 1* (Ottawa: Information Canada, 1973).

14. National Petroleum Council, *U.S. Energy Outlook: An Initial Appraisal, 1971-1985*, Volumes 1-3 (Washington, D.C.: National Petroleum Council, 1971).

15. The Department of Energy, Mines and Resources, *op. cit.*

5.
LEGAL CONSTRAINTS ON
PETROLEUM POLICY OPTIONS
IN NORTHERN CANADA

Andrew R. Thompson and Michael Crommelin

A. THE CANADA OIL AND GAS LAND REGULATIONS

The first problem in the management of oil and gas resources is the allocation of rights between the public and private sectors and the choice among competing private applicants. In the Northern Territories the federal government dealt with this issue not through legislation but by exercising its power to make regulations afforded by the Territorial Lands Act, which states:

> The Governor-in-Council may make regulations for the leasing of mining rights in, under or upon territorial lands and the payment of royalties therefore . . . 1

The regulations which are currently applicable, the Canada Oil and Gas Land Regulations,[2] were promulgated in 1961 at a time when there was little interest in oil and gas in the north. Twelve years later, it is clear that the situation has radically altered. It is therefore important to examine both the nature and the extent of the rights that have been allocated under the regulations to date,

TABLE 1

Region	Potential Reserves	Area of Sedimentary Basin	Area of Current Permits
Yukon and Northwest Territories	35 billion bbl. oil 210 trillion cu. ft. gas	160 million acres	135 million acres
Arctic Islands (inclu. offshore)	50 billion bbl. oil 300 trillion cu. ft. gas	384 million acres	305 million acres
East Coast Offshore —Grand Banks —Scotian Shelf —Labrador —Gulf of St. Lawrence	32 billion bbl. oil 189 trillion cu. ft. gas (to water depth of 200 metres)	217 million acres (to water depth of 200 metres)	300 million acres
Hudson Bay	not available	128 million acres	71 million acres
West Coast	not available	31 million acres	16 million acres

Various sources.

and to see how far these rights operate as a constraint upon future policy choices for Canada.

Figure 1 shows that in Canada north of 60°, the region administered by the Department of Indian Affairs and Northern Development, there are 544 million acres of sedimentary lands that are potentially petroleum producing. The Canadian Petroleum Association estimates that the potential reserves of this region are 85 billion barrels of recoverable crude oil and 510 trillion cubic feet (t.c.f.) of natural gas. By way of comparison, Canada's offshore regions on the east and west coasts and in Hudson Bay, which are administered by the Department of Energy, Mines and Resources, are estimated to include 376 million acres of potential petrolum producing sediments containing 32 billion barrels of recoverable crude oil and 189 t.c.f. of natural gas, up to a water depth of 200 metres.

Of the northern acreage underlain by sediments, some 440 million acres, or 80 per cent of the total, are currently under permit. The remaining 20 per cent consists of the marginal areas where sediments are thin, exploration results have been disappointing, or in the case of offshore areas, water depths are such that exploitation is not feasible under current technology. A quick glance at a permit map reveals that all the areas of interest, the Mackenzie Delta, the Arctic Islands and the northern offshore sedimentary basins are blanketed with permits.

Figure 1 shows when the northern acreage was taken under permits. It is notable that the total amount was modest until 1968 when the Prudhoe Bay oil discovery was made. In fact, acreage under permits actually declined in the mainland Northwest Territories after the new regulations were introduced in 1961. By 1970, however, the Prudhoe Bay take-off had escalated to the point where almost no acreage remained for disposition.

Figure 2 shows how the regulations divide the northern areas into three zones with increasingly generous terms being afforded as one moves into the more remote northern regions and into the offshore.

Figure 3 shows how the work requirement, which varies from $2.65 per acre to $2.90 per acre depending upon the region in which the permit is located, is spread out over the permit term, so that on a present worth basis the cost of work requirements to

FIGURE 1

ACREAGE HELD UNDER OIL & GAS PERMIT

YUKON TERRITORY AND NORTHWEST TERRITORIES

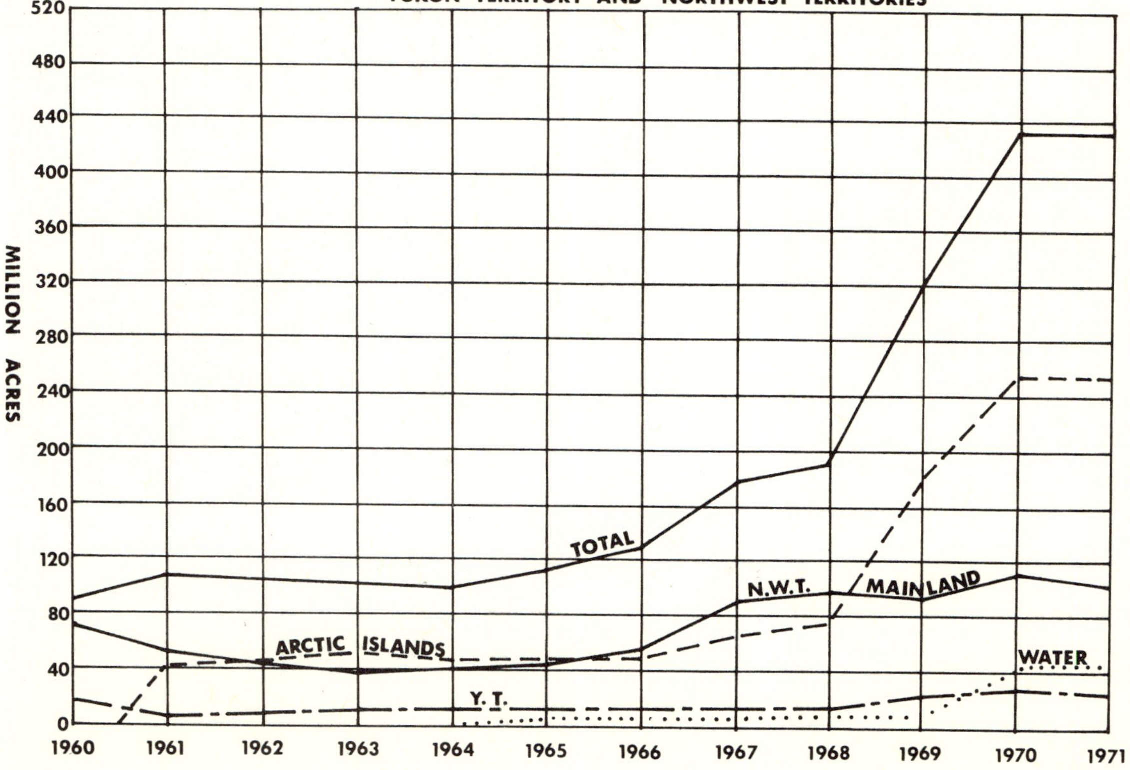

hold a permit for 12 years in the most northerly region is little more than $1.00 per acre. Furthermore, being expressed as work requirements, these amounts can be earned by offsetting exploration expenditures incurred on the permit or on those with which it is grouped. There are no other costs of holding acreage apart from a $250.00 filing fee. The result is that the great bulk of permits taken out in the north from 1968 onwards can be maintained in force until the early 1980's under these very favourable terms.

The next question is what rights are given by the permit. First, it should be realized that basic exploration by seismic techniques can be carried out anywhere by the holder of a exploration licence, as distinguished from a permit. The licence holder can conduct his exploration, short of drilling a well, even on lands held under permit or lease by someone else. The permit is therefore a second-stage disposition. It gives the holder the right to carry his exploration to the point of drilling a well. Section 35(1) of the Canada Oil and Gas Regulations also provides that:

> Subject to these regulations, a permittee has the exclusive option to obtain an oil and gas lease for the Canada lands described in his exploratory permit.

The permittee can choose when to exercise this option to lease except that if he discovers oil in commercial quantities, the government can give him notice requiring him to take out his lease within one year. In result, where oil has not been discovered, the permits, which have lasted for 12 years and take us into the 1980's, can be converted into leases which will last for a further 21 years. We are now in the next century. Should oil be discovered in commercial quantities during the time of this lease, the lessee has the right to have the lease reissued for a further period of 21 years at the same royalty rates as applied under the original lease. The initial permits and leases thereunder can last without discovery of oil or gas into the next century.

THE CROWN RESERVE SYSTEM

Initially, the permittee is not entitled to lease all of his acreage. Rather, he is entitled to only 50 per cent, with the remaining half being surrendered to the Crown for disposal by tender. This practice of lease selection on half of the acreage, with the remaining half becoming a Crown reserve which can be sold at

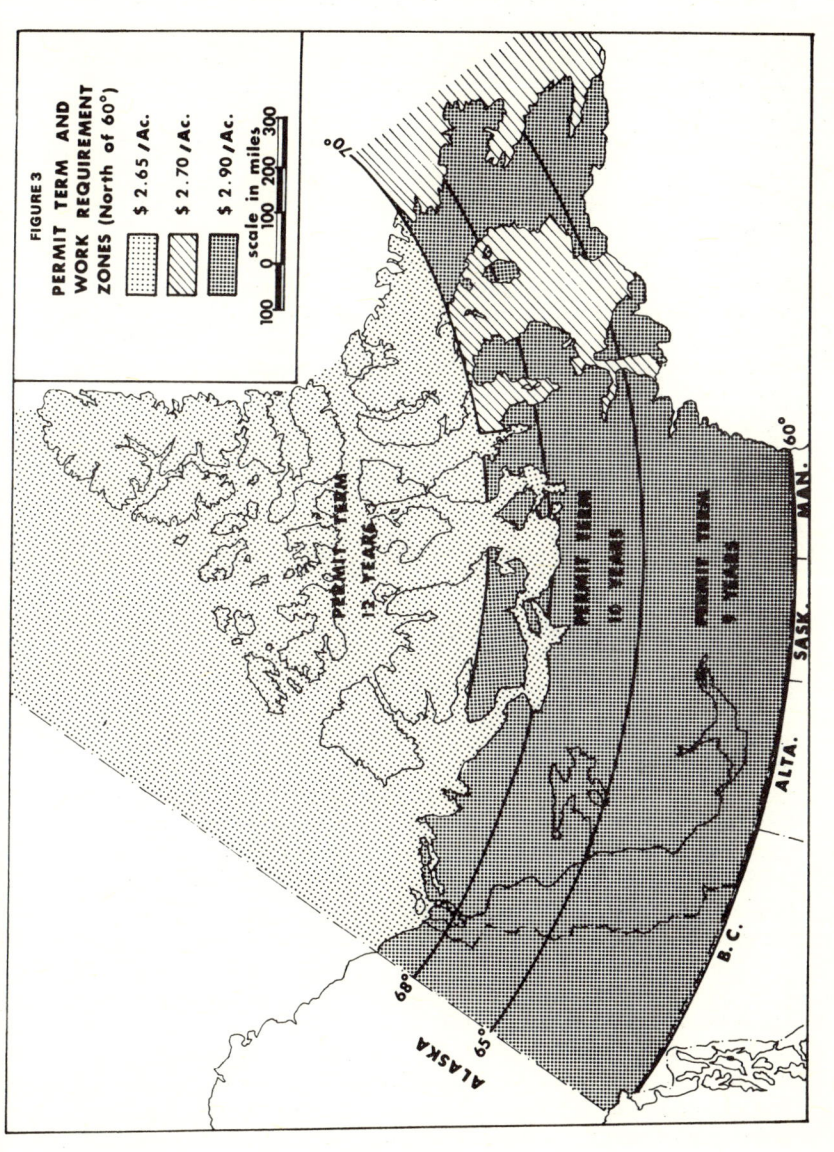

FIGURE 3

PERMIT TERM AND WORK REQUIREMENT ZONES (North of 60°)

$ 2.65 /Ac.

$ 2.70 /Ac.

$ 2.90 /Ac.

scale in miles

100   0   100   200   300

PERMIT TERM 12 YEARS

PERMIT TERM 10 YEARS

PERMIT TERM 9 YEARS

ALASKA

B.C.   ALTA.   SASK.   MAN.

60°

55°

68°

70°

auction, is well established in Canada. Since the 1947 oil discovery at Leduc, this practice has enriched the western oil producing provinces, most notably Alberta, whose lease sales have produced in excess of $1,500,000,000. The theory is that the permittee will have the opportunity to select what he believes to be the most favourable acreage, but at least one-half of the permit area will return to the Crown so that the Crown may share in the rewards which the permittee's exploration efforts have produced. This is considered to be a due return for the various exploration incentives, including the free acquisition of the permit in the first place, which the government has afforded to the permittee.

This system is to be contrasted with the leasing system in Alaska and in the United States offshore where there are no permits, and all leases must be acquired directly by purchase. Canadians have been impressed by news of lease sales which in 1968 paid the United States Federal Treasury $603,000,000 for 71 leases covering 363,000 acres in the Santa Barbara Channel, in 1969 netted the State of Alaska $900,000,000 for 164 leases covering 413,000 acres on the North Slope, and in September and December 1972 paid the United States Treasury $2,551,000,000 for 178 leases comprising 830,000 acres off the Louisiana coast. In 1971 the United Kingdom government conducted a limited experiment in the sale of production rights by inviting tenders for 15 production licences in the North Sea, for which successful bidders paid more than $89,000,000.

OIL AND GAS LAND ORDER NO. 1-1961

In Canada, however, the federal government in 1961 in effect gave away the northern and offshore Crown reserves. The Oil and Gas Land Order No. 1-1961 stated that a permittee, after selecting his leases, could apply for leases on the remaining area which the regulations required him to surrender to the Crown. This meant that the permittee could take leases on the entire acreage covered by his permit without any payment. There was, of course, a penalty. It was that the additional leases selected over and above the initial 50 per cent would bear an additional royalty rate. However, this additional royalty rate was not in any sense an adequate substitute for the bonus payments which the Crown might otherwise have received. In such a large area as the northern territories, both production and cost characteristics of fields will vary considerably from place to place, making it

FIGURE 3
YUKON TERRITORY – NORTHWEST TERRITORIES
PERMIT TERMS AND DEPOSIT REQUIREMENTS – PER ACRE

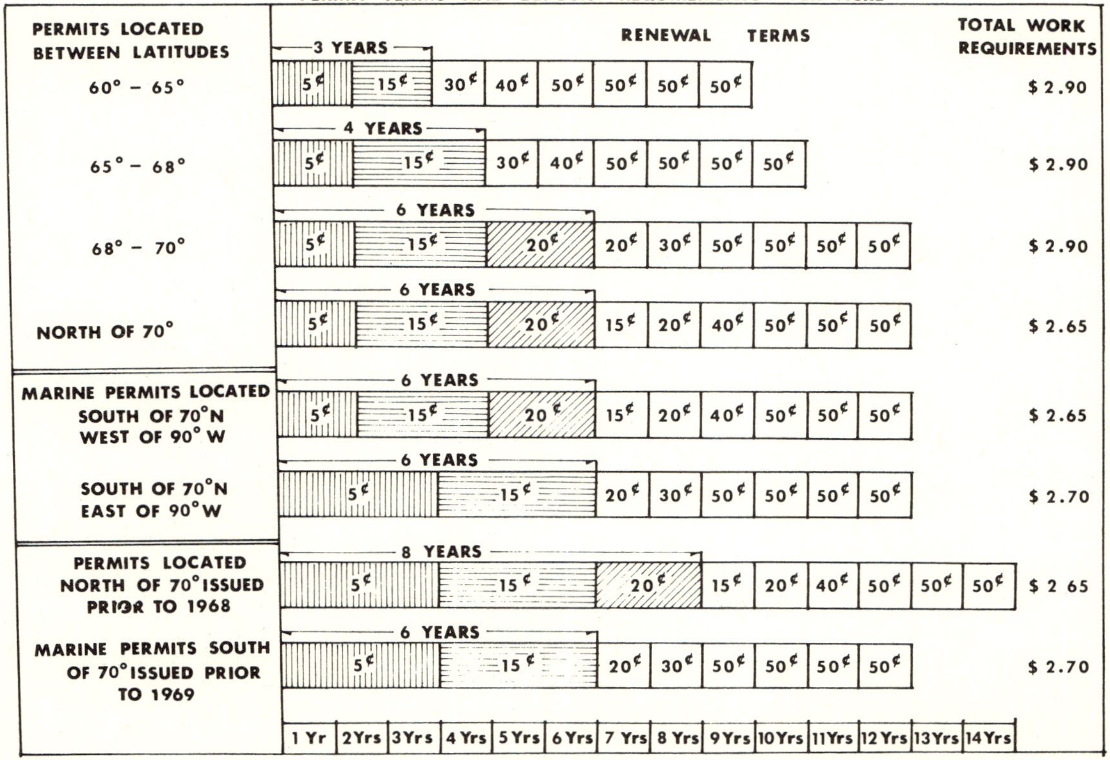
PERMITS LOCATED BETWEEN LATITUDES
RENEWAL TERMS
TOTAL WORK REQUIREMENTS
60° – 65°
3 YEARS
5¢ 15¢ 30¢ 40¢ 50¢ 50¢ 50¢ 50¢
$ 2.90
65° – 68°
4 YEARS
5¢ 15¢ 30¢ 40¢ 50¢ 50¢ 50¢ 50¢
$ 2.90
68° – 70°
6 YEARS
5¢ 15¢ 20¢ 20¢ 30¢ 50¢ 50¢ 50¢ 50¢
$ 2.90
NORTH OF 70°
6 YEARS
5¢ 15¢ 20¢ 15¢ 20¢ 40¢ 50¢ 50¢ 50¢
$ 2.65
MARINE PERMITS LOCATED SOUTH OF 70°N WEST OF 90° W
6 YEARS
5¢ 15¢ 20¢ 15¢ 20¢ 40¢ 50¢ 50¢ 50¢
$ 2.65
SOUTH OF 70°N EAST OF 90°W
6 YEARS
5¢ 15¢ 20¢ 30¢ 50¢ 50¢ 50¢ 50¢
$ 2.70
PERMITS LOCATED NORTH OF 70° ISSUED PRIOR TO 1968
8 YEARS
5¢ 15¢ 20¢ 15¢ 20¢ 40¢ 50¢ 50¢ 50¢
$ 2 65
MARINE PERMITS SOUTH OF 70° ISSUED PRIOR TO 1969
6 YEARS
5¢ 15¢ 20¢ 30¢ 50¢ 50¢ 50¢ 50¢
$ 2.70
1 Yr 2Yrs 3Yrs 4 Yrs 5 Yrs 6 Yrs 7 Yrs 8 Yrs 9Yrs 10Yrs 11Yrs 12 Yrs 13Yrs 14Yrs

impossible to design a single system of royalty rates which would recover what might otherwise be received through bonus bidding. Furthermore, since the permittee was given the choice of whether or not to select additional leases, he would only do so if he expected the additional royalty payments to be less than the price for which the leases could be acquired at a sale. Accordingly, the option to take additional leases operated exclusively in his favour.

There is also a technical reason why this system of additional royalties paid only on the additional leases is undesirable. It creates the equivalent situation to fragmented ownership of an oil and gas reservoir and imposes on the government's oil and gas administrators an almost unmanageable burden. The problem is that individual oil and gas wells will in many cases produce petroleum substances from underneath both an original lease and an additional lease. Because reservoir conditions are not uniform, it will be necessary to allocate this production between the two in order to apply the different royalty rates. A simple acreage basis of allocation is unfair. Consequently, complicated "tract factors" must be calculated for each lease. These are notoriously the subject of difficult and prolonged negotiations when oil companies and lessors are seeking unitization in a province like Alberta where divided ownership of a reservoir is not uncommon. Under the additional leasing system similar negotiations will have to be carried out between the oil company representatives and government administrators. Unlike the case when such negotiations are carried out between two or more companies, each armed with reservoir data, all of the information about the well will have been generated by the company and the government administrator will find himself in the difficult position of having to rely on the permittee to inform him what he is bargaining about!

THE ROYALTY RATE

Figure 4 shows the royalty rates – the percentage of wellhead price of oil paid to the government – applicable in the north and offshore and in the other producing provinces. It is notable that the rate in the federal areas is substantially lower than that elsewhere. The 5 per cent for the first three years is unprecedented. Furthermore, the 10 per cent thereafter is to be compared with a 16 2/3 per cent rate in the United States offshore

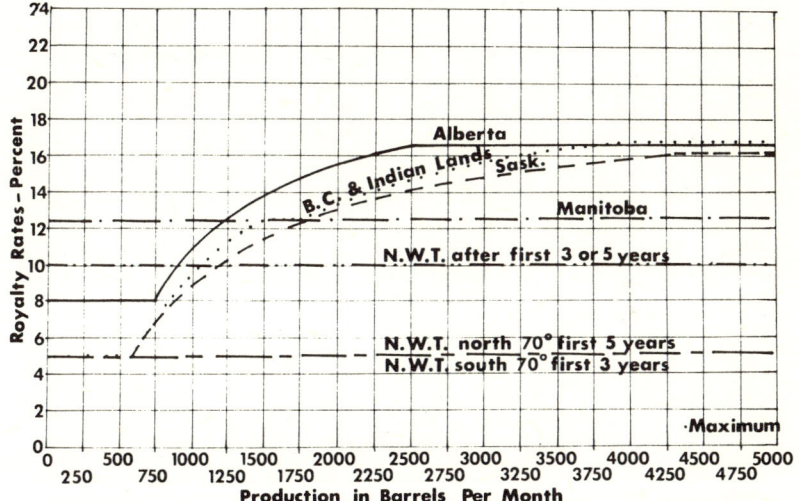

FIGURE 4

COMPARISON OF CROWN ROYALTY RATES – WE'TERN CANADA
CRUDE OIL
June 1st, 964

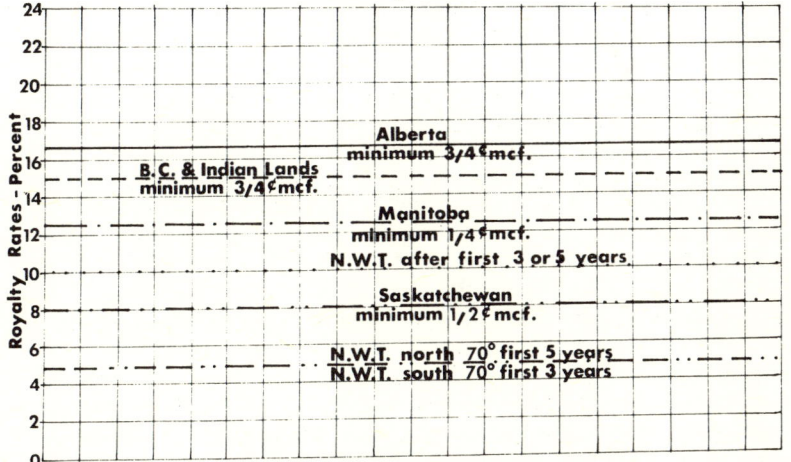

COMPARISON OF CROWN ROYALTY RATES – WESTERN CANADA
NATURAL OR RESIDUE GAS
June 1st, 1964

Value Calculated Back at the Wellhead

areas and a 20 per cent effective rate in Alaska. The figures for Alberta and British Columbia shown in Figure 4 are no longer current. In Alberta the Lougheed government has increased the high end of the variable scale to 25 per cent for oil, and the average is now 21 per cent. In his recent budget speech the Premier of British Columbia announced that the royalty rates for oil in that province would be increased to achieve a scale of between 10 and 40 per cent.

The loss of revenue to the federal government from these extremely low royalty rates will be large indeed, as two examples will illustrate. The first is a gas pipeline from the Mackenzie Delta to southern Canada and U.S. markets, with a throughput of 4.5 billion cubic feet per day of gas obtained in equal proportions from the Alaska North Slope and the Mackenzie Delta, according to the Consortium's proposal described in Chapter 2 of this volume. If we postulate a well-head value of 30 cents per thousand cubic feet in both places (a figure mentioned in reports of gas prepayment contracts for the Delta), the royalties payable to the Canadian government during the first three years of production would be approximately $36 million as compared with $122 million payable to the State of Alaska. Thereafter, the Canadian government would receive $24 million per year and the Alaskan government $40.65 million per year. Over a ten-year period the Canadian gas revenues would be $202 million less than those in Alaska.[3]

The second example is an oil pipeline from the Mackenzie Delta to the same markets with a capacity of 1.5 million barrels per day. We assume a well-head value for the oil of $2.00 per barrel. Royalties payable to the Canadian government during the first three years of production would be $157.5 million. The same oil produced in northern Alberta would generate royalties of $787.5 million for the provincial government at the 25 per cent rate. After the first three years the federal government would collect $105 million per year, as compared with $262.5 million for the Alberta government. The ten-year difference in oil revenues would be $1,732.5 million in favour of Alberta.[4]

These figures, large as they are, refer to only two of a number of possible oil and gas developments in the north and offshore. The Mackenzie Delta may well prove capable of production greatly exceeding that used in the above examples. Very large gas discoveries have already been made in the Arctic Islands, and hopes remain high for oil in this region.

B. THE NEED FOR CHANGE

Considering that northern oil and gas resources are publicly-owned, the ultimate goal in their management must be the public benefit, whether viewed from the interests of northern residents or of Canadians as a whole, or of some mix of these interests. When resources are developed by private entrepreneurs many public benefits and costs are not taken into account in the ordinary way of market decisions. For example, important public benefits may be improvement in a country's balance of trade position, increased employment of labour and capital, and regional economic development. Public costs are such things as environmental degradation and interference with people's chosen way of life. It is also the government's responsibility to make choices between the present and the future when resource development decisions are taken, because the discount rates employed by private entrepreneurs are unlikely to coincide with the time preference of society as a whole. The government has a role as arbiter of social values, and this role must have a place in any management regime for oil and gas resources. The regime should remain responsive to government policy decisions which can take account of changing social values.

A further aspect of the public benefit derived from northern oil and gas resources is the government revenues obtained when these resources are exploited. The appropriate objective for the government to seek here is to maximize the present value of these revenues. Only in this way will the distribution problem be solved and the benefits from resource development accrue to the resource owners, the public.

The two objectives identified here, responsiveness of the management regime and government revenue maximization, are interdependent to a degree, in that the control retained by the government to ensure responsiveness may give rise to uncertainty in the private sector and thereby discourage investment. It is therefore a matter of reasonable balance to devise a management system which will allow the required measure of responsiveness to changing circumstances without unduly affecting private investment.

We may now ask whether the Canada Oil and Gas Land Regulations establish a regime for management of northern oil and gas resources in the public interest. Our answer can only be no.

If the public interest is tested by the extent to which the system is responsive to government policy it fails because it has tied the government's hands for excessively long periods to terms and conditions that were laid down when the regulations were made a decade ago. This assertion requires further documentation.

At the very initiation of the exploration process, government control is non-existent because the present regulations provide for a "free-entry" system. Companies have been able to obtain permits merely by filing applications with the Department of Indian Affairs and Northern Development. Under this free-entry system, no influence is exercised by the government over the total acreage under permit at any time, over the particular areas in which oil and gas operations are to take place, over the nature and ownership of the companies requiring permits, or over the amount and kind of exploration and development work to be done. In contrast, offshore rights have been obtainable in the U.S., the U.K. and Australia only following a government invitation for applications in respect of designated areas, and the influence of the government in selecting the successful applicant and reviewing his proposed work programme has been considerable.

The regulations also give rights that are clearly excessive in duration. As we have previously shown, most northern permits and all offshore permits may remain in force for a total of 12 years, when the holder has the right to select leases with a further term of 21 years. The lessee then is given the right to have the lease re-issued for a new term of 21 years from the date when commercial production is obtained. In the extreme case, should production be obtained in the final year of the basic term, the permit, initial lease and re-issued lease combine to provide a duration of 53 years from the date when the original permit was granted. The significance of this extraordinarily long duration lies in the fact that basic terms and conditions, including royalty rates, lie beyond government control. The closest approximation to this long duration is to be found in the 60- and 99-year concessions that the Arab sheiks granted in their unenlightened period prior to World War II.

Nor do the regulations promote a turnover of areas held under permit which would enable the government to regain control from time to time. Whereas the Australian system requires relinquishment by a permittee of at least half of his permit area after six years, and then half again after each succeeding period of five years, the Canadian regulations provide for no such relinquishment. Furthermore, an Australian permit holder may obtain production rights only in respect of a commercial discovery of oil or gas, whereas his Canadian counterpart can lease without the necessity of a discovery. Under the United States offshore system turnover is assured since there are no permits in the first place, and leases expire after only five years unless they are prolonged by discovery and commercial production.

Finally, if the public interest is tested by the measure of revenue received by the government when a decision is made in favour of exploitation of oil and gas, the regulations again fail. We have shown earlier that Canada's royalty rates are lower than those in comparable jurisdictions and that further major revenue losses will occur if the bonus bid system is given up in exchange for higher royalty rates on additional leases.

We think the case for change is established. The Canadian regulations have permitted almost all of the northern sedimentary basins to be tied up for excessively long terms without any requirements of discovery of oil and gas, at a cost to the companies which is minimal as compared with that in other countries. By these regulations the federal government has given up the opportunity to manage northern oil and gas resources for the public benefit.

C. THE OPPORTUNITIES FOR CHANGE

In the light of the Alaska sale of Prudhoe Bay oil leases in 1969, the federal government became aware that Oil and Gas Land Order No. 1-1961 (which allowed existing permit holders to apply for additional leases over the Crown reserves) was an improvident arrangement so far as the public interest was concerned. Therefore in 1970, in an announcement that startled the oil industry, the government revoked the Order. In taking this stand, the government relied on the fact that the language of the Order was merely discretionary and inappropriate for conferring a right on the permittee to obtain additional leases. Since at that time few permittees had applied for leases, the effect of revoking the Order was substantially to restore the original system whereby Crown reserves could be disposed of by competitive sale. The industry reaction was abrupt and vigorous. It argued that the Order gave the permittee a "vested right" to obtain the additional leases and that the government could not change the rules mid-stream. In the context of this confrontation, the government announced that it would undertake a major rewriting of the oil and gas leasing regulations.

Oilweek, the Canadian trade journal, reported on October 9, 1972, that this review was then almost complete. It stated:

> It has been learned that all work on the regulations has been completed by the respective department staffs. Industry has been in contact with the government on these changes and several "frank, entirely open and understanding" discussions have been held.

The same opportunity for frank and open contribution to the new leasing system has not been afforded to other Canadians. The draft changes have been discussed with the industry on a confidential basis and have not reached the news media or other interested persons. Were it not for the "colonial" status of the northern territories, these changes would come before Parliament in a statute. But as regulations, there is not even a requirement that they be exposed to debate before a parliamentary committee. Nor will the territorial councils have a say.

The importance of this process is underlined by the fact that *Oilweek* also reports that the government *will retain the present system* for the northern acreage now under permits. This will make both the review and the proposed changes virtually meaningless, for, as we stated previously, the prospective oil and gas regions are blanketed with permits. In addition, it appears that the industry has succeeded in its lobbying for reinstatement of Oil and Gas Land Order No. 1-1961, whereby a permittee could apply for additional leases. In fact, industry may have even enhanced its position because it is proposed that the government's *discretion* to grant additional leases from the Crown reserves be converted into a *right* on the part of the permittee to acquire them. The only concessions which the permittee would make is that the additional leases would now last for only ten years without discovery instead of 21 years (still twice as long as U.S. offshore leases), and the additional royalties imposed on production from these leases would be revised upward. Since the permittee would retain the choice of whether or not to apply for additional leases, this latter concession is not of great importance.

The reason given by the government for not changing the regulations governing existing permits is that permit holders have what lawyers refer to as "vested rights", which entitle the holders to retain their permits together with the terms and conditions, including royalty rates, which existed *at the time the permits were first issued.* We challenge this reasoning. First, it is absolutely clear that the terms and conditions of existing permits can be changed if Parliament so decides. In Canada the concept of "vested rights" means only that when changes are made, compensation should be paid if the changes have either taken away property rights or result in a breach of contract by the government, unless Parliament specifically legislates to remove any right to compensation. There have been many examples throughout the world where existing rights of concession holders have been taken away subject to payment of compensation. Such is currently happening in Saudi Arabia where the state is taking up to 50 per cent of the rights of concession holders. In Canada we do not hesitate to take property rights where it is in the public interest to do so. Oil companies themselves expropriate private property for well sites and pipeline rights-of-way. But in Canada as a matter of public policy we would take the step of taking away private rights of the kind given by the oil permits only in exceptional circumstances. We believe that the extraordinary changes in national consciousness about resources, and particularly about energy, have produced exceptional circumstances of the kind that justify changes. But our opinion in this respect is not important. What is important is that the Canadian public should have the opportunity to make a judgement about this matter.

We also question the legal basis, itself, of the argument that changes made in the regulations affecting existing permits would violate "vested rights". There is an element of doubt as to the nature of the legal rights acquired under a permit. Some argue that the permittee has no property or contract rights, but merely a statutory privilege that can be withdrawn. Further, if the permit holders do have "vested rights", these rights are not nearly so extensive as to entitle them to retain the permits for the full duration of 12 years and to keep their leases in force for the full period of 21 years thereafter. There are no less than four situations during the lifetime of the permits and leases when changes could be made without interfering with vested rights.

First, since the acquiring of additional leases under Oil and Gas Land Order No. 1-1961 was not a right but a discretionary matter, and because this Order has now been revoked, the federal government could say to the oil companies that it will not grant leases in excess of the original 50 per cent selection unless the permittee agrees to exchange his original permit and *all* leases granted thereunder for rights under an entirely new system more attuned to the public interest.

Second, since renewal of permits in the 8th, 9th, 10th, 11the and 12th years is also discretionary (section 38), the government could insist in the 8th year that a permittee who has not "gone to lease" must surrender his rights in exchange for rights under the new system.

Third, whenever a permittee or lessee applies to transfer an interest in his permit or lease, a transaction for which government consent is required (sections 72-77), the government could insist that these rights be exchanged for rights under the new system.

Fourth, in the fourth year of all leases, the government could give the lessee the option of drilling a well on each of his leases or accepting rights under the new system (section 89).

If legal reasons do not stand in the way of making changes that will affect the terms and conditions of existing permits and leases, are there economic considerations that militate against such changes? The argument is made that the uncertainty produced by such changes will deter the entrepreneur from making further investments in exploration for oil and gas in the north and offshore even if the new terms offered are by themselves attractive, because he will have lost confidence in the system ("the investment climate"). This loss of confidence may be analysed as a loss of certitude that the terms and conditions won't be further changed to the detriment any new investment. In familiar terms, one does not deal for long with someone who does not keep his part of the bargain. We agree that there is force in this argument, though this force is greatly diminished where a government has a monopoly over resources that are needed as Canada's will be in the next decade.

Also, there is a further factor which engineers and scientists will readily understand. The Canadian leasing system has, in our opinion, got so far out of equilibrium with the systems in other oil producing regions that it has become politically unstable. The pressure for change will, in our opinion, become intolerable once oil production begins and Canadians become aware that they have forgone bonus bids and adequate royalty rates as compared with other regions. Oilmen are aware of this build-up of pressure for change; some of them will even now be discounting it when evaluating their investment choices. In these circumstances, even from an economic standpoint, it is better for the government to make the necessary changes now to provide a system that investors can confidently expect will endure for the foreseeable future. It is like the proposals of seismologists to trigger minor shocks along major fault systems by underground explosions so that a major earthquake will not occur in the future. We suggest that a minor shock to investors now is preferable to a major shake-up in a few years time.

CONCLUSION

There is ample precedent both within Canada and without for government action of the type recommended here. A current example in the Middle East is provided by the recent statement of the Shah of Iran informing the oil consortium operating in that country that it could either retain its rights under the 1954 oil agreement until they expire in 1979, when they would expect no favour, but would have to "stand in a long queue to buy Iran's oil without any privileges over the other customers," or they could now accept a new arrangement whereby Iran would take over the producing rights and the consortium would obtain long-term purchase rights of 20 or 25 years at preferred prices and discounts. In Alberta, the Lougheed government last year told oil and gas lessees that they could either accept a new top royalty rate of 25 per cent instead of the 16 2/3 per cent stated in the leases, or be subject to a new tax on oil reserves. It is reported that most lessees will accept the new royalty rates.

Neither can it be said that this type of government action is a new or radical innovation. In 1962 the Government of Alberta achieved a revision of its oil leasing legislation whereby the former 21-year leases were phased out and ten-year leases were substituted along with other more onerous requirements. The industry accepted these changes because their need was clearly stated and the new terms were not unfair. We believe that if the federal government were to state its determination to bring about substantial revision in the system, the oil industry would cooperate towards phasing in a new system which would strike a balance between serving the public interest and honouring the investment which the industry has made.

6.
IMPACT OF AN
ARCTIC PIPELINE
ON NORTHERN NATIVES

Stuart Jamieson

The potential impact of the northern gas pipeline on the native people along the proposed route appears on the surface to be a separate question from its impact on Canada as a whole. The two, however, are related in important ways.

The net gains that the pipeline development would generate for the people of the North are not likely to be proportional to the benefits that would accrue to other Canadians. Indeed, northerners could be made worse off while the rest of Canada benefitted, or (for instance, through increased subsidies), the reverse could happen. In addition, of course, all Canadians could gain or lose (a question that is analyzed in Chapter 10 of this volume).

Apart from inputs of materials, machinery and other supplies, most of which will come from the rest of Canada or from abroad, a disproportionate share of the direct expenditures for construction are likely to be made in the North. This seems likely, since about 900 of the 2,500 total miles of the proposed pipeline lie in the Yukon and Northwest Territories, where the most difficult terrain and climatic conditions will be encountered.

These expenditures will generate corresponding labour and business incomes in the North. However, because most of the labour used in the construction and operation of the pipeline will have to be brought in from the South, only a fraction of the income benefits will accrue to native and other residents. The North is also likely to suffer a disproportionate share of the social and environmental damage or costs involved in pipeline construction and operation, as a report by the federal Department of Finance has pointed out.[1] Whether the proposed project will, on balance, improve or worsen the economic and social "welfare" or "well-being" of the native population of the North is exceedingly difficult to analyze. Adequate information about some highly important and relevant data is lacking, and much of such data is difficult to analyse with any degree of precision.

The question of net gains and losses in income and well-being for the native people in the North is not just a matter of the economics of pipeline construction. It also depends upon federal government policies affecting the way it is built: what precautions are taken, what regulations are imposed to minimize social and environmental damage, what programs and policies are formulated to deal with the social problems that the pipeline and related developments are likely to create, and how effectively such policies are carried out. In addition, the impact of such development will depend upon the federal government's fiscal policies affecting revenues and expenditures in the North.

The North is, and has been, heavily subsidized by the rest of Canada. The governmental budget for the Northwest Territory currently has been about $120 million per annum, and for the Yukon Territory, about $50 million per annum, for a total population of hardly more than 40,000. Some 85 per cent or more of these funds comprise federal government expenditures. A disproportionate share of these, relative to the total population, consists of direct monetary payments and provision of various goods and services to the native population. In return, the natives produce few exports of cash value. They are, in brief, the recipients of tens of millions of dollars annually, in cash or real income, primarily from the federal government.

If the northern territories were in a position to extract the full economic value of their gas resources, and of the land required for the pipeline, their potential income would contribute substantially toward balancing their accounts with Canada as a whole. But, as the analysis in Chapter 10 suggests, most of the resource value is likely to accrue as profits to producing companies and to consumers.[2] As a result of its conspicuously generous tax and royalty concessions to oil companies operating in the North, the Canadian government's share of the potential gains is negligible. The Yukon and Northwest territorial governments are expected to receive extra revenues of only a few million altogether, mainly from fuel and property taxes.

Unlike those in Alaska, therefore, the natives in Canada's North will not receive large cash settlements, land grants and annual incomes derived from rents, royalties and taxes. Their benefits are expected to be almost entirely in the form of new job and income opportunities. But only a fraction of the jobs created will go to the natives.

The net gain to the native population will depend to a large extent on how the federal government will adjust its heavy subsidies to northerners as development takes place. If, to take an extreme example, the government reduced its expenditures in the North, dollar-for-dollar, in direct proportion to any gains that accrue to northerners as income and taxes, then there would be no net dollar gain to the resident population. And, in view of probable social and environmental damage, the whole project would result in a *net loss* in income and welfare for the native population. Only if the government were to maintain or increase its expenditures in the region (or at least reduce them by less than any gain in labour income) could one safely conclude that the native population would stand to gain any benefit from the project.

Purely monetary estimates of gains and losses are, of course, insufficient to assess the impact of the pipeline on the native people. Traditionally, real income per capita has been the standard measure of social welfare, in Western industrial society, and economic growth has been coterminous with gains in general well-being. This view has come under increasing attack in recent years, on the grounds that the quality of life as well as the quantity of output has to be considered in assessing the benefits of economic growth. Such problems as congestion, environmental damage, family breakdown, alcoholism and drug addiction, mental illness, and other such phenomena have generally increased with economic growth and rising per capita income in most developed countries. Their incidence seems all the more likely to rise (as they already have) in areas such as the McKenzie Valley, where there are the added strains of ethnic and cultural change and conflict as the native population attempts to adjust to a new way of life being imposed upon them. These problems impose

additional economic costs of governmental and charitable assistance designed to deal with them. But, more important, they
result in incalculable social and psychological strains and hardships, all of which should properly be charged as "costs," against
the gains in real income from such projects as a gas pipeline.

Moreover, despite drastic change and disruption brought by
increasing contact with white society, native Indians and Eskimos
in the North have managed to retain some strong roots in their
aboriginal cultures. If the well-being of the native population is
to have a high priority, such roots should be viewed as "values"
that are important in providing the basis for a separate and
distinct sense of identity. The erosion or outright destruction of
such values as a consequence of the gas pipeline should then be
viewed as an additional "cost" to be assessed against any gains
in real income that may be derived from the project.

On balance, therefore, if construction of a McKenzie Valley
pipeline is to enhance the well-being of the native population, it
will have to generate a large (perhaps very large) economic gain
to compensate for unquantifiable psycho-socio-cultural damage
that the project is likely to impose on these people. This gain
would have to cover both the costs of providing adequate protection of the social and natural environment, as well as rehabilita-
tion measures to deal with unavoidable damage, and increased
expenditures to support and expand cultural, leisure-time, recreational and other facilities and services for the native population.
These seem likely to be needed to provide natives with the
means for pursuing meaningful lives that will better enable them
to resist the social disruption and disorganization that a large-
scale project like the pipeline is likely to bring.

A. THE IMPACT OF WHITE SOCIETY ON NATIVE PEOPLE

A popular objection to the proposed project to build a gas
pipeline up the McKenzie Valley is that it threatens to destroy
the "native way of life." What is this "way of life"? In what
respects is it likely to be destroyed?

Broad portrayals of the social history and character of northern Indians and Eskimos are now a familiar part of Canadian
literature. They depict cultural norms that, in various respects,
differ sharply from those of the prevailing white North American

society. Native economies and cultures were based on a harsh climate, a difficult terrain and limited resources, and they utilized ingenious but limited technologies. These economies could support only a small population over a large territory at a very precarious margin of subsistence. Such environmental and technological constraints gave rise to certain values and behavior-patterns essential to survival. Among these are the following:

(i) Human association in nomadic small groups or "bands." These groups, based on ties of family and kinship, carried on hunting, fishing, trapping and other resource-gathering activities over a radius of 50-100 miles. This type of social organization contrasts with the increasing concentration of people in large and permanent urban centers characteristic of Western industrial society.

(ii) Short time perspective. In precarious subsistence economies most people cannot afford to devote much time, effort or resources to the accumulation of capital or other durable goods, acquisition of complex or esoteric knowledge or techniques, or other long-term objectives. It takes full-time effort to live from day-to-day.

(iii) The sharing ethic. Where the outputs of hunting, fishing and other food-gathering activities are highly variable and uncertain, and require the joint efforts of several individuals and families, the imperatives of survival call for sharing the proceeds of one's labour rather than appropriating them as one's own, for greater personal comfort or status in competition with others.

(iv) The sexual division of labour. The position of women in aboriginal cultures based on bare subsistence economies is often pictured as "depressed," "subordinate," or "menial" by comparison with prevailing white North American norms. The cultural carry-over is such that it seems to render native women particularly vulnerable to exploitation, sexual and otherwise, in white-dominated communities and situations.

Contact with white society through a few industries – trapping, whaling and (particularly since World War II) military and other government activities – has brought rapid and drastic changes in the Eskimo and Indian way of life, and rapid erosion of their traditional economies and cultures. Concentration in settlements with special facilities and services has been associated with a decline in the traditional mobility involved in fishing, trapping and hunting. Increasingly, resources within easy reach of settlements have been over-utilized and depleted, while those more distant have been under-utilized.[3] Governmental health and welfare services, perquisites, and other income supplements have been conducive to a rapid rate of population increase. Some scholars and observers in the North maintain that it is already over-populated in view of the meagre capacity of the northern economy to support a large and growing number and variety of opportunities for productive employment.[4]

Extensive contact of natives with the prevailing white society has had extremely disruptive and demoralizing results. Native cultures, in brief, tend to be overwhelmed. Whites, having control of essential administrative machinery and services, as well as capital, technologies, markets and outside contacts, dominate positions of money and power in the economic and political structure. Natives, to the extent that they are brought into or "integrated" in the structure, are relegated largely to unskilled, menial or dependent positions. A more or less rigid caste system has thus tended to develop in which the natives are locked into positions of inferior income and status. They are thus vulnerable to a denigrating self-image, with costly social and psychological results.[5]

In terms of orthodox criteria it can be argued that, despite all this, the native people have gained economically. The average per capita real income from all sources (i.e. from jobs, welfare, education and health services, etc. taken together) is undoubtedly higher than it was, or would have been, in the aboriginal state. The native cultures, based on bare subsistence economies of nomadic hunting and fishing, comprised a harsh and demanding way of life that involved innumerable and intense hardships, strains and conflicts. The modern white influence has made life easier, more affluent, more secure and less demanding. Nevertheless, in view of the new and serious psycho-socio-cultural strains and conflicts that have ensued, it would be difficult to demonstrate that it has created happier or more meaningful lives for the native population.

This broad picture applies to most of the native Indian population over Canada as a whole, and not just to Indians and Eskimos in the Northern territories. Despite unprecedented economic expansion and new job opportunities in many parts of Canada since the war, and the new and expanded educational, training and job placement programs of the Indian Affairs and Manpower Departments, relatively few natives (in the south or in the north) have become effectively "integrated" into the white economy. The limited studies available indicate that unemployment (including "disguised" unemployment) ranges from almost one-half of all employable Indians in British Columbia to two-thirds or more in some Prairie and Maritime provinces.[6]

The main barriers to effective integration of natives appear to be social or cultural rather than technical:

(i) Subtle difficulties confront individual workers from a visible minority in established work environments. Informal groups (cliques, factions and friendship groups) effectively enforce various norms of behaviour that are difficult for the outsider to appreciate and learn.

(ii) Where the worker is geographically separated from his home village, family and kinsmen, strains and anxieties cause a high turnover.

(iii) Where the worker can bring his family to the community in which his job is located, the major barriers often lie in the cultural difficulties that his family (particularly his wife) faces in fitting into the community.[7] Unsuitable standards, from the majority white point of view, of household maintenance and management, child care, dress and comportment, and unfamiliarity with the ways of the white community generally, tend to invoke condemnation or ostracism from neighbours. The native housewife, facing social isolation, frustration and loneliness, tension and strain within the family, often resorts to alcohol. Pressure on the husband tends to be the same, with such consequences as excessive drinking, absenteeism and suspension or dismissal.

(iv) Where a native settlement is located within commuting distance of the place of employment, all the above strains may be absent, but the "sharing ethic" referred to earlier may generate equal pressure. Regularly employed and relatively well-paid native workers are often pressured to share their bounty with less fortunate friends or kinsmen. Refusal to do so can lead to hostility and social isolation, and again, internal strains in the family.

This, in rough outline, has been the prevailing situation, even before large-scale gas exploration and development. As to the impact of large-scale activity that pipeline construction will bring, we can only make an eclectic prediction. If such development were unplanned or poorly planned it could make the present situation worse by accelerating the adverse trends outlined above. On the other hand, through careful planning and administration, it might be possible to take advantage of the new opportunities to substantially improve the lives of native peoples in the North, and arrest or reverse some of the adverse trends.

B. THE CURRENT ECONOMIC CONDITION OF NORTHERN NATIVES

The present economic condition of the native population along the route for the proposed gas pipeline is extremely marginal at a low, poverty level. So far there has been limited direct involvement in, or direct economic benefit from, the new urban industrial-commercial economy (a large part of which represents government facilities and services).

Natives comprise more than one-half the population but account for less than one-third of the total labour force in the area, and this fraction, on the average, is employed only a few months of the year. All told, some 1517 (or almost two-thirds of the total native labour force) are concentrated in the traditional activities of hunting, trapping and fishing, which yield conspicuously low returns. The gross value of output from these activities (including the estimated value of domestic consumption as well as sales) was assessed in 1969-1970 at $1,108,000 for natives in the McKenzie Valley and $43,000 for the Old Crow area in the Yukon – or hardly more than $750 per annum, on the average, for each of the 1517 participants. (This contrasts to the total output of some $99,521,000 from the non-renewable resource industries, mainly lead and zinc mining and smelting.) This total represents only a very small fraction of the total labour income from all sources of almost $87 million in the area. Only a handful of natives are employed in such fields as mining and smelting (2.3 per cent of the native labour force, and about 10 per cent of all workers in the industry), managerial, technical or skilled jobs. Most employment of natives in fields outside of the traditional resource industries lies in unskilled, casual or seasonal jobs in the categories of "service and recreation" and "unspecified labourers." In comparison with the resource industries, however, the roughly one-third of the native labour force engaged in wage employment accounts for more than 67 per cent of the total native "earned income" and about 58 per cent of native income from all sources: in aggregate terms, some $2,465,382 from this source, as against $1,150,000 from hunting, trapping and fishing.

Serious underemployment prevails among the native work force. Hunting, trapping and fishing are highly seasonal, accounting on the average for only a few months employment each year. Among those engaged in wage employment, some 47 per cent worked from 4 to 24 weeks per annum during 1970. "Unearned income," in the form of welfare and other transfer payments, in the aggregate provide a larger total income to the native population than do hunting, trapping and fishing, and accounts altogether for 20-25 per cent or more of income from all sources. The average or "mean" incomes per family in the McKenzie district are estimated at $2568 for native Indians, $4643 for Eskimos and $5136 for Métis. In view of the high prices of food and other imports, these represent poverty levels in comparison with an average family income of $9186 for non-natives.[8]

On the face of it, there is little of economic value remaining in the so-called "native way of life." The two-thirds of the native labour force engaged in "traditional" activities of hunting, trapping and fishing earn, in the aggregate, less than one-third of the total "earned" income of natives, while more than two-thirds of their total income is received by the less than one-third engaged in wage employment in non-renewable resource industries and trades. But even the latter employment is largely unskilled, seasonal and casual, involving extended periods of unemployment for most workers and requiring, therefore, substantial subsidization through "transfer payments" of various kinds to provide minimum levels of subsistence.

The level of participation of the native population in traditional activities of hunting, trapping and fishing has declined over the past couple of decades for reasons outlined earlier. The decrease has occurred more in the form of shorter periods of active participation, and a lesser intensity of effort, than in actual numbers of people involved. Orthodox economic criteria would indicate a clear gain in economic efficiency and well-being of the native population if larger numbers moved entirely from the traditional activities of hunting, trapping and fishing and into wage employment. This would presumably apply as long as there remained any measurable differential in average per capita "real" earnings and in "psychic satisfaction" between the two sets of activities. Such a conclusion seems open to serious qualification, however. The apparently declining interest and involvement of natives in their traditional activities is not due solely to the lower real income available as compared to alternative wage employment. For the latter is, in effect, heavily subsidized and made more attractive by federal government policy. The latter provides, in the main settlements, a variety of facilities and services in education, health, housing, social welfare, and other fields. These encourage a long-term flow of population from hinterland areas. The incomes generated by personnel employed in the provision of such services, in turn, support numerous private business enterprises which likewise create wage and salaried employment. They are similarly concentrated in the main settlements.

The fact that almost two-thirds of the native labour force continues to participate in the traditional resource-based activities despite the apparently superior attractions of wage employment in the main centres, together with the fact that only a minority of native workers have been willing and able to commit themselves to full-time wage employment, suggests that the former continue to have a "cultural value." If such activities were as heavily subsidized, and provided with some of the same advantages as wage employment has, they might well induce a greater degree of participation from native workers.

C. IMPACT OF A GAS PIPELINE ON THE NATIVE PEOPLE

Only very rough estimates can be made of the possible impact of the proposed gas pipeline upon the native economy portrayed above. One could attempt to predict from previous experience, of course, but this has obvious limitations.

A number of large-scale projects, similar in some respects to the proposed pipeline, have been carried out in various northern areas of Canada over the past two decades, such as the Peace River Dam in northern British Columbia; the Trans-mountain Pipeline from northwestern Alberta to southwest British Columbia; and the development of large-scale mining and smelting operations in northern Manitoba. All of these involved large-scale investments, concentrating economic activities in remote areas that contain sizeable native Indian populations. Their impact upon the natives appears to have incurred high social costs, and (indirectly) economic costs, that have not been adequately assessed in orthodox economic terms.

The typical course of events has been something like the following: Capital-intensive development generates a boom in an area during the construction phase. Large numbers of single transient workers are brought in – a type who tend to be prolific spenders, particularly on liquor and women. Natives are temporarily drawn into the new economy. Men for a time can earn "big money" (by comparison with traditional hunting, fishing or trapping) on temporary construction jobs, and this tends to be spent rapidly on expensive durable goods and on alcohol. Women are drawn into the economy in various service jobs or into prostitution. Most of the new income is rapidly drained out of the area for imported goods and services, and little in the way of new long-term employment or income is generated locally. After the construction phase is over, relatively few job opportunities remain.

Most natives have been unable to take advantage of the few new long-term job opportunities. Many are also unable or unwilling to return to their traditional hunting, fishing or trapping activities after the taste of "high life," and become partially or wholly idle and dependent on welfare. Social disorganization and demoralization are manifested in destructive forms, such as family breakdown, neglect of children, malnutrition and excessive drinking. Permanent new economic and social costs are thus imposed on the community or region as a price for economic development.

The above picture perhaps exaggerates the problems that might result from a large-scale pipeline. Critics, however, maintain that, by the very size and scale of the operations that are proposed, their effects are likely to be far more disrupting and damaging to the native population than any other single project that has been undertaken in Canada. On the face of it, the introduction of thousands of additional workers and billions of dollars of new machinery and equipment in the space of a few years, into a region having only a few thousand people now obtaining their livelihoods from a fragile economy and environment, is likely to have an extremely damaging impact.

Supporters of northern pipeline development, on the other hand maintain that the potentially damaging effects of such development can be avoided or minimized through proper advance planning, careful regulation and control; and that the economic gains to be derived from the pipeline and related developments are far greater than any achieved from the limited development in the north to date. These economic benefits will be of such a magnitude as to make possible, one way or another, substantial improvements in well-being and a more satisfactory way of life for residents of the North, particularly those of native background.

Supporters of the project also foresee the possibility of more long-term employment opportunities for native residents. Barring major new gas and oil discoveries in the North, however, such predictions have to be viewed as pure conjecture as far as the current McKenzie gas pipeline proposal is concerned.

The federal government is supporting construction of the pipeline, while attempting to mitigate the problems that a project of this kind tends to create. To this end, the Minister of Indian Affairs and Northern Development has announced certain policies and "guidelines" for the construction and operation of pipelines and related facilities.[9] These are intended primarily to minimize damage to the physical and biological environment, to prevent disruption of established communities and traditional ways of life in the North, and to maximize, for Northern residents (particularly those of native stock) new income and employment opportunities that the project will make available. But whether these broad guidelines are adequate in terms of the objectives they are designed to achieve is an open question that requires further examination. Such organizations as the Environment Protection Board, the Canadian Arctic Resources Committee and the Inuit Taparisat of Canada, have criticized the guidelines as inadequate.[10]

The policy of DIAND, as suggested in the "guidelines," is to give priority of employment on the pipeline and related projects to native workers wherever possible, by assuring special training and counselling services and other means. A much-quoted target is to achieve and maintain the employment of 1000 native workers. At the high rates of pay that will prevail on pipeline work, this could mean an average of $1000-1500 a month. Construction work on the pipeline will, it is anticipated, have to be concentrated largely during the four winter months, and be carried on over a period of two-and-a-half to three years. In total then, such work could yield to the native population $1-1.5 million a month or $3-4.5 million per annum, for a total of $9-13.5 million altogether during the entire construction phase of the pipeline project. *If* the 1000 workers employed represented a net addition to the native participating labour force, *or if* the seasonal work on the pipeline project could be perfectly dovetailed with seasonal hunting, trapping and fishing activities and with casual wage jobs now being engaged in, such that the present native labour force were employed an extra four months or more each year, *then* the estimated $3-4.5 million per annum would constitute a very substantial net gain in cash income. Indeed, it might exceed the total present earnings which, as noted earlier, amounted to $2,465,382 from wage employment and $1,150,000 from hunting, trapping and fishing. (As against this "net gain," however, there would presumably be a reduction in welfare and some other transfer payments that natives now receive to sustain them during periods of unemployment.)

It seems unlikely, however, that such neat dovetailing of jobs will be possible. Most of the natives recruited for work on the pipeline probably will be drawn from the (approximately) 2400 now engaged, on at least a part-time basis, in other wage employment and in the traditional resource-based industries. (Indeed, it might well be argued that the new recruits will comprise the more resourceful, skilled, productive, and/or "acculturated" workers from these other fields). If one were to assume the extreme case, such that all 1000 pipeline workers were recruited from their present employments, and abandoned these latter entirely, then "other things being equal," this would mean a reduction of 40 per cent in the 2400 workers so engaged and a corresponding reduction in their total income by some $1,440,000. This would reduce the net gain in income to the native population from pipeline employment to a range of roughly $1.56-3 million per annum during the three years of construction. A more realistic estimate would probably lie somewhere between these broad extremes of $1.56 to $3 million per annum, or a total of from $4.68 to $9 million for the three-year construction phase as a whole.

It would seem unrealistically optimistic, however, to view the "opportunity costs" of the native population as comprising the sacrifice of alternative employment and income for pipeline work *during the three-year construction phase alone.* To take the most pessimistic view (judging from the impact of some previous large-scale construction "booms" on native populations) many native workers employed for two or three years in high-paid pipeline jobs would be unable or unwilling to revert to low-paid menial or labouring jobs, or to hunting, fishing and trapping when the construction phase of the project reached completion. If such were to occur, the situation might well represent a permanent new level of costs. Not only would there be the long-term sacrifice of present employment and income. There would also be the additional public expenditures that would be incurred in maintaining new recruits to the ranks of unemployed dependents, as well as the new and additional outlays incurred in dealing with such problems as increased alcoholism and family disorganization or breakdown. In the longer run, over five, ten or more years, again assuming "other things being equal," such additional costs might well exceed the additional income benefits from temporary pipeline work.

All this, it may be objected, presents too narrow and pessimistic a picture. The proposed pipeline project will, it is maintained, have a "multiplier effect" that will generate other job and income opportunities for the native population. There will be some 5-6000 other workers in addition to the 1000 natives, earning in the aggregate $30-45 million per annum over the three-year period. Expenditures from this source alone in the McKenzie district will generate millions of dollars of additional labour and business income, and hundreds, possibly thousands, of new jobs.

The "multiplier effect" is likely to be somewhat limited in the North however. Its deficit position in relation to the rest of Canada, and its extreme dependence upon outside imports of all kinds, create large and rapid "leakages" of income. Furthermore, the ability of native Indians and Eskimos to take advantage of such new job and income opportunities seems extremely limited. Out of a total labour income recently amounting to more than $88 million per annum, as noted earlier, the share received by the native labour force amounts to a mere $3-4 million. Native workers account for barely 10 per cent of the total labour force employed in mining and smelting, clerical and sales work, and transportation and communication, and far smaller percentages in managerial, professional, technical and various types of skilled and self-employed occupations, in the McKenzie district. Unless a large and intensive program of training native workers were undertaken beforehand by DIAND or other agencies, there is little reason to expect that native workers would account for any larger percentages than these in the new jobs created in "spin-offs" from the pipeline project.

Nor do the longer term prospects appear, on the face of it, to be more favourable. By far the major expenditures upon, and incomes generated from, the pipeline project, are expected to occur during the 2½ to 3 year construction phase. Thereafter the operation and maintenance of the pipeline is expected to require hardly more than 200 permanent employees. Most of the income generated during this latter period will accrue to, and be expended by, gas and pipeline companies as well as by the federal government in the south. Unless there were continued exploration on a large scale, and major new oil and gas discoveries in the McKenzie Valley or nearby areas – and these cannot be assumed to arise from the gas pipeline in itself – there is little reason to expect that the project will generate in the long run, directly or indirectly, any great number of new job and income opportunities for the resident native population.

D. AN OPTIMUM PROGRAM FOR THE NATIVE PEOPLE

On balance there seems to be, at best, a limited "surplus" of cash or real income potentially accruing to the native people from the proposed gas pipeline. It appears insufficient to compensate for the many unknown, and in terms of dollars incalculable, risks of damage to the socio-cultural and ecological environment of the Yukon and McKenzie territories. If the federal government is determined to continue sanctioning the project in the face of these uncertainties, logically it should be prepared to invest large sums of money in comprehensive and carefully planned programs. These should be designed, not only to protect and prepare the native population against potential damage and disruption. They should also have the more positive objectives of reviving and strengthening the economic base and cultural independence of the native peoples, to enable them to achieve a more meaningful and satisfactory way of life and thus be better able to resist some of the more damaging effects of the pipeline project. This objective would call for such policies as the following, among others:

(i) Maximum aid, in training and counselling services, in credit, and the like, for those natives and their dependents who wish to take advantage of the job and business opportunities that the pipeline and related developments will generate;

(ii) Special effort and expenditure, in financial and technical aid, for those of native background who are unable or unwilling to become committed to long-term wage or salaried employment and who prefer to depend upon traditional resource-based activities; and

(iii) For those who are unable or unwilling to engage in these two broad categories of employment, subsidization and encouragement of cultural and leisure-time activities that hold some promise of being a meaningful alternative to idleness and demoralization.

Needless to say, a program of this kind will require the utmost cooperation, cousultation with and active participation from the native people of the North, in research and planning as well as in administration. How else are we to interpret their concepts of optimum "welfare" or "well being"? The natives, through their more-or-less representative organizations and spokesmen, have been expressing growing concern about the implicatious of the proposed pipeline, and growing dissatisfaction about programs that are formulated on their behalf and imposed upon them by the federal government, which call merely for "comments".

Suggestions such as these may seem rather far removed from the question raised by the title of this chapter. But until various alternatives are discussed and experimented with as matters of practical policy, there are few tangible answers one can offer regarding the potential impact of the proposed gas pipeline upon native people in the North.

FOOTNOTES

1. This report was "leaked" to the *Canadian Forum* and was published in the June-July, 1973 issue of that journal.

2. See also, Milton Moore, "Canada's Perverse Oil and Gas Policies," in *Canadian Forum* (June-July, 1973), pp. 27-29.

3. John R. Wolforth, *The McKenzie Delta – Its Economic Base and Development* (Ottawa: Dept. Indian Affairs and Northern Developtent, MDRP, 1970).

4. Peter J. Usher, *The Bankslanders – Economy and Ecology of a Frontier Trapping Community* (Ottawa: Dept. of Indian Affairs and Northern Development, NSRG 71:1-3, 1971).

5. J. M. Lubart, *Psychodynamic Problems of Adaptation of McKenzie Delta Eskimos* (Ottawa; Dept. Indian Affairs and Northern Development, MDRP 7, 1970).

6. W. T. Stanbury, D. B. Fields and D. Stevenson, "Unemployment and Labour Participation Rates of B. C. Indians Living off Reserves," *Manpower Review Pacific Region* (April-May-June, 1972); H. B. Hawthorn, ed., *A Survey of Contemporary Indians of Canada*, Part I, Chap. 3 (Ottawa: Indian Affairs Branch, 1967).

7. S. M. Jamieson and H. B. Hawthorn, *The Role of Native People in Industrial Development in Northern Manitoba, 1960-1975* (Winnipeg: Committee on Manitoba's Economic Future, 1962); P. J. Usher, *Economic Basis and Resource Use of the Coppermine-Holman Region, N.W.T.* (Ottawa: DIAND, 1965).

8. The statistics in the above section were derived mainly from J. R. Wolforth, *The Evaluation and Economy of the Delta Community* (Ottawa: Northern Science Research Group, DIAND, 1971); C. Y. Kuo, *A Study of Income Distribution in the McKenzie District of Northern Canada* (Ottawa: Environmental Social Group, DIAND, 1973); and J. Palmer, *Social Accounts for the North* (Ottawa: Environmental-Social Group, DIAND, Interim Paper No. 3, 1973). M.P.S. Associates Ltd., "Regional Impact of a Northern Gas Pipeline – Impact of Pipelines on Traditional

Activities of Hunter-Trappers in the Territories," *Report* (Ottawa: DIAND, 1973). DIAND – *North of 60 – Mines and Minerals* (Ottawa, 1972). DIAND – Water, Forest and Land Division, *Report* (Ottawa, 1972). Statistics Canada, *Fur Production* (Ottawa: Catalogue 23-207, 1973).

9. "Expanded Guidelines for Northern Pipelines," *Report* Tabled in House of Commons, June 28, 1972 by the Honorable Jean Chrétien (Department of Indian Affairs and Northern Development, Ottawa, 1972).

10. *Briefs* from these organizations were submitted to the Director, Environment-Social Program, Northern Pipelines, Ottawa, dated in the order listed above, as follows: December 22, 1972; November 17, 1972; December 4, 1972.

7.
ENVIRONMENTAL CONSIDERATIONS
IN NORTHERN RESOURCE DEVELOPMENT

Everett B. Peterson

Despite today's public concern over environment and despite
suggestions that there is now an opportunity for thoughtful
appraisal of all issues, it is concluded in this paper that the
ecological side effects and trade-offs involving renewable components of the environment will not be major considerations in the
decision on whether or when gas and oil pipelines should be built
from the north. The accumulated knowledge about northern
ecological conditions and responses will find application instead
in the avoidance of problem areas during such projects, in the
design of measures to lessen environmental side effects or to
ensure the safety of man-made structures, and in the development of stipulations for long-term management of renewable
resources of the north. Such a conclusion calls for explanation
because acceptance of the stated conclusion suggests that the
entire subject of "environmental impact assessment" in advance
of proposed projects is less important in decision-making than
the public has been led to believe. Carried to its extreme, the
conclusion may lead one to ask, "Why carry out advance ecological studies at all?" The remainder of this paper will focus on this

question by pointing out both the value and the limitations of applying northern ecological information to a project such as a gas trunkline.

A. THE ROLE FOR ECOLOGICAL STUDIES IN ADVANCE OF MAJOR PROJECTS

Kreith's recent analysis of a random sample of 200 environmental impact statements in the United States indicated that in no cases had listed adverse environmental effects resulted in the proposed project being abandoned. Of 127 statements which listed adverse environmental effects, a total of 214 alternatives were listed, all of which were rejected: 130 were rejected for economic reasons; 47 were rejected for environmental reasons; (i.e. the alternative would do more harm than the proposed action); and 37 were rejected because of engineering problems[1]. There may be various explanations for these statistics, but there seems to be a tendency for actions to proceed essentially unchanged, despite predictions of environmental degradation. Why should this be so at a time when there appears to be so much interest in the social management of technological consequences?

1. SEPARATING THE ROLES OF SCIENTISTS AND THE PUBLIC IN TECHNOLOGY ASSESSMENT

One reason for the apparently conservative role of environmental impact statements is that we are in the midst of a search for better ways of technology assessment. The objective of this search is to "scrutinize the interactions, side effects, by-products, spillovers, and trade-offs among several developing technologies or between a new technology and society at large and the environment."[2] That we are not yet successful in this search is evident, but there does seem to be agreement that the novel aspect is the emphasis on adding to the cost-benefit equation the second-order effects which, in the long run, may affect society more deeply than the intended primary effects.

When ecological data are gathered to assist in technology assessment, there is one immediate problem because of the current tendency by many people to indiscriminately combine as "environmental considerations" moral or political questions and scientific questions. An example of the moral environmental question is one that takes the form, "What sort of environment do we want?" and is typified by Dr. Donald Chant's contribution to the recent *Financial Post* feature, "Our Costly Cleanup"[3]; an example of the scientific environmental question could be one that seeks to determine recovery time of a tundra area after distrubance and is typified by the work of Dr. Ross MacKay[4]. This paper will deal only with what we are here calling the scientific environmental questions. The problem for environmental impact assessment arises when this distinction is not made. For example, we are told that the north will be developed without upsetting natural balances. This is misleading because, as Feinberg[5] has pointed out, there are an infinite number of possible "balances of nature" rather than a unique one. If there are man-made disturbances, all components of the natural system will adjust to new values. The new values may be more or less desirable than the old ones, but this depends on the ethical or aesthetic criteria by which they are judged, and so it is a moral or political question rather than a scientific one. Therefore, when a scientist speaks about vehicle ruts on the tundra there is no doubt that he is describing a "change" to that ecosystem but it can be debated whether that change is "damage"; if the scientist chooses to define the change as damage, rather than leaving that definition to society, the most that he or she can do is to define "damage" arbitrarily[6]. Environmental impact statements written by scientists are sometimes criticized because they do not say whether or not the proposed project should be approved or rejected on environmental grounds. Such a criticism should never be made if one accepts the distinction in the role of scientists and the public.

There seems to be agreement that the scientist, in advance ecological studies, should document the following: the part of the environment affected; how it is affected; and the likelihood, timing, magnitude, duration, diffusion, source and controllability of side effects. Knowing these things, the scientist then has a role as a consultant and should be required to recommend alternatives, but it is up to a much larger segment of society to decide if the predicted environmental impact is acceptable and which of the alternatives is best. Having argued that it is not up to the scientists alone to decide whether or not a particular proposed project should go ahead, when we refer back to the statistics that show a lack of correlation between documentation of adverse environmental effects and abandonment of proposed projects, one conclusion is that public apathy must be a major cause of bad environmental management when it occurs[7].

2. THE PROBLEM OF EVALUATING ENVIRONMENTAL COSTS

The problem of evaluating environmental intangibles[8] and the difficulty in measuring non-marketed values in comparable economic terms are obvious deterrents to effective consideration of ecological data during technology assessment. Those who plea for determination of the economic value of the arctic environment[9] have argued, using wildlife habitat as an example, that the habitat's worth must be established because otherwise money may be spent on protecting wildlife habitat for a group of people who might not utilize the wildlife and who might be better off if the funds were spent directly for housing, education or cultural purposes. Although there have been some discussions of the opportunity costs of projects that lead to loss of whole environments,[10] and some recent development of methods to evaluate non-priced recreation resources,[11] there is generally no economic measurement of the value of wilderness[12] or the value of rare scientific research materials, such as the archeological site a few miles from proposed pipelines in the northern Yukon where there is evidence of man's occupation at least 10,000 years earlier than previously known for any part of Canada.[13] As a result of our incapability of measuring such benefits, Fisher and co-workers[10] suggest that we are left with no choice but to ask what the present value of preserving an area would need to be to equal or exceed the present value of the development alternatives. Although there have been a number of proposed projects in Western Canada in which environmental counter-arguments resulted in abandonment of the project plans, in a project as large as the proposed oil or gas pipelines from the north, the present value of the development alternative is generally assumed to be too great to be outweighed by the benefits that would go with outright preservation of the area.

One interesting attempt to compare the value of preserving a northern area with the alternative of allowing development to take place is the analysis by Naysmith.[14] Considering the net benefits accruing to the community of Old Crow in the northern Yukon from various degrees of regulatory control on land use, Naysmith examined three sources of income; (a) geophysical exploration employment, (b) pipeline maintenance employment, and (c) hunting and trapping. The value of the land to the Old Crow Indians in terms of exploration employment was estimated at $120,000 for a 5-year period; the value of the land for providing pipeline maintenance employment was estimated at $700,000 for a 20-year period. In both of these cases the value of the land would not increase by application of more restrictive regulations. The situation is different with the third alternative where the land is taken as wildlife habitat. Taking the average annual dollar income represented by the harvesting of caribou and fur-bearers at $70,000, the value of the land, provided its productive capacity is not impaired, could be assessed at $1,400,000 over a 20-year period. This is more than the value accruing to the local residents from the two industrial operations combined. Naysmith appropriately points out that the important difference here is that, unlike the geophysical operations and the pipeline maintenance, increasingly restrictive land use regulations correspondingly increase the value of the land as wildlife habitat. If no restrictions are imposed, the value of the land in terms of a hunting and trapping economy could be reduced to zero. On the other hand, rigid regulatory control of land use operations could maintain the value of the habitat at the full $1,400,000.[15]

This example from the Yukon indicates that we may not be too far from a capacity of applying ecological criteria to decisions on whether or when to proceed with projects such as a northern gas trunkline. However, work to date, while mentioning a concern for environmental side-effects of various oil and gas transportation proposals in the Mackenzie River valley, typically does not incorporate "environmental costs" into the cost-benefit comparisons of several alternatives.[16] Even the recent work by Cicchetti[17] is quick to point out that environmental effects are uncertain and usually difficult to quantify in monetary terms. We are left then with the question: How great do the environmental damages have to be in order to make the choice between preservation or development? Casting this question in terms of the proposed Trans-Alaska oil pipeline, Cicchetti has estimated that the present economic value of this Alaska pipeline to the nation, at a 10 per cent discount rate, is about three billion to six billion dollars or almost 15 billion dollars undiscounted. Under such circumstances, a decision to totally forego development of north slope oil would mean that the nation judged the environmental damage to Alaska and to west coast waterways and ports to be greater than or equal to this amount. If these figures be only approximately true for a gas pipeline from the north, it appears that environmental considerations, in a strictly analytical comparison, would not match the economic return from utilization of the fuel resource; this does not deny, however, the importance of environmental considerations as a political force.

3. THE TIME-SCALE PROBLEM IN ECOLOGICAL STUDIES

A third barrier to the effective consideration of ecological information in decisions to approve or reject projects such as a northern gas pipeline is the long time required for many kinds of environmental studies. Ecologists' pleas for more time to carry out necessary data collection are well publicized. Surprisingly, we often overlook that the time required to obtain sound environmental data on the behavior of rivers and ice, population changes of wildlife species, responses to surface disturbance, or rates of recovery of various ecosystems is probably no greater than the time required to explore and prove up a major oil and gas field in the north; in both cases a decade seems to be a reasonable time requirement. So if there is a time problem for ecological studies, it is often just a problem of phasing the studies to coincide with industrial plans. It is a fact that comprehensive environmental studies did not begin in phase with, or in advance of, geophysical exploration in the north and this is one of the reasons why environmental data are not advanced to the point that we might wish them to be for any trade-off analyses that enter into decisions on whether or when the last phase of oil and gas development (transportation) is to take place. Known sedimentary basins in the north are currently about 80 per cent committed to exploration permits. These permits would neither have been granted nor sought had there not been an underlying assumption that oil and gas finds (above certain threshold amounts) would be brought to market. If environmental costs are to be a real factor in calculating whether or when such fossil fuels should be removed from the north, then environmental considerations should have been a part of the assumptions and decisions at the time of acreage dispositions. Instead of this, environmental considerations have entered the picture only as we approach the time of commitment for transportation of the resource

The weakness of placing a considerable environmental research effort on only the later phases of a proposed development needs no elaboration. Instead, we should seek ways to ensure that necessary background ecological studies take place in phase with the necessary exploration on the part of industry. In regard to the observation that environmental studies did not begin ten years ago when there was a commitment to explore for gas and oil in the north, some will suggest that there was not an environmental awareness among the public at that time. While this may be part of the answer, there is still today an apparent difficulty for early initiation of project-related environmental studies in our research planning organizations. An example will demonstrate this point. Maybin,[18] speaking on behalf of the Canadian Gas Association, has described development of an industry producing synthetic gas from coal in the 1980's. Each individual plant producing such gas would involve a very substantial coal mining venture, because a standard size plant is expected to use six million tons of coal per year. This is larger than any single coal mining operation in Canada at present and Maybin judges that there is a prospect for 100 or more of these plants in the next decade in Canada. Some of these would involve surface mining. With such advance warning, surely this is an opportunity for the necessary environmental impact studies to get started with a reasonable lead time; yet there are no obvious moves in this direction. The point here is that there will be scientists still documenting the *actual* effects of a gas pipeline from the north, let us say 10 years from now, when research planners will have to assign their personnel to *begin* to predict the environmental consequences of the next phase of gas development which is likely to be the production of synthetic gas from coal reserves and which, in a decade, will already be upon us.

In summary, we are dealing, at best, with environmental impact *prediction* and not environmental impact *assessment*. Actual assessment of the environmental effects of such a project can be done only from documentation during construction, operation, and abandonment of the facility. Accuracy of the prediction is limited by the relatively long sampling time required to establish probabilities for various side-effects, and by the fact that many environmental changes may result either in damages that last for a very long time or which are delayed in their onset.[19]

If ecological side effects and trade-offs involving renewable components of the environment will not be important elements in the decision on whether or when gas or oil pipelines should be built from the north, then it is necessary to ask, "Why carry out any advance pipeline-related environmental studies?" One reason is that for some kinds of "environmental studies" there is legitimacy to the claims that "the protection of the environment is necessary to assure the safety of the pipeline,"[20]that "the objectives of the conservationists and the pipeline owner are in fact the same"[21] or that "good engineering and environmental protection are synonymous."[9] These statements are true for many physical components of northern ecosystems such as near-surface groundwater, surface water, unstable slopes, riverbanks, or ice-rich soils where minimizing disruption to these parts of the environment is obviously in the interest of pipeline safety; but to extend to all of the biological components of northern ecosystems the slogan "engineering that is good for the pipeline is also good for the environment" is, of course, nonsense.

Though biologically oriented studies may be of little direct value in ensuring a safe pipeline, they can find application in at least four important contexts. First, insofar as native land claims are based on their use of renewable resources of the land, information on wildlife and fishery populations, movement patterns, habitat productivity and carrying capacity is necessary for domentation of traditional land use and for recognition of biologically sustainable yields in any future attempts to guarantee these traditional land uses. Second, the management and protection of northern wildlife and fisheries species requires detailed knowledge of their habitat requirements, life cycle stages, and responses to harvest or other disturbances. Third, the successful enforcement of land-use regulations depends ultimately upon a wide knowledge of responses of vegetation to surface disturbance, recovery times of terrestrial and freshwater ecosystems, and biologically sound methods of reclamation. Fourth, international treaty commitments, as in the case of migratory waterfowl, depend very much upon protective measures for those habitat and life cycle stages that are provided by the north. Questions of international equity are included here, as in the case of salmon produced in the northern Yukon. Such productivities in Canadian waters can be a factor in international negotiation of the annual salmon harvest by Canadian fishermen in offshore areas. Unlike many of the environmental components that are not amenable to economic evaluation, this example, involving a run that can reach 250,000 fish in some years,[22] could be readily considered in trade-off calculations in the event that this renewable resource were threatened by construction and operation of a pipeline.

B. IMPORTANT ENVIRONMENTAL CONSIDERATIONS FOR NORTHERN PIPELINES

Technical details of the ecological side effects of northern petroleum development have been reviewed in a number of recent articles.[23] Most of these articles focus on two ecological aspects of the Arctic which are already well known in the public media: (a) the biological simplicity of the ecosystem in terms of the limited number of species and therefore the greater potential danger in eliminating the key species within the food chain; and, (b) the presence of permafrost and the potential for severe terrain modification should it be ice-rich. To date (April 1973) there has not been public identification of any specific environmental component that would be irretrievably lost or unacceptably damaged by the proposed pipeline projects. As expected, there are frequent assertions by the potential applicants for gas and oil pipelines that they are convinced there will not be undue damage to the environment. Other analyses[24] have also indicated that there is little doubt that a pipeline can be built through the Mackenzie Valley without major disruption to the land-based animals and plants and to the land itself, providing appropriate and usually expensive precautions are taken.

Many investigators believe that the greatest environmental uncertainty for a northern pipeline is the problem of delineating high ice-content soils. In some cases there are definite surface indicators but, as Hill[9] has pointed out, for every known location of sub-surface ice there is another unknown location that could not be detected other than by drilling. There is a need for delineation of buried ice bodies along proposed pipeline routes and for any other areas within the permafrost zone scheduled for development. The main reason for identifying this as the major problem is that perennially frozen ground underlies approximately half of the land area of Canada[25] and there is no sure method to predict the location of all underground forms of ice. There is no need to document here the well-publicized phenomena of melting, subsidence and slumping after disturbance of the surface layer of organic material. Only the possible magnitude of these phenomena need be indicated by the documented example of a gully 23 feet wide and 8 feet deep that had eroded in four years along a bulldozed seismic line west of the Mackenzie Delta.[26] The implications of these potential erosional features for pipeline facilities are obvious. Adding to the difficulty is the fact that gravel, which is important for the type of construction that calls for insulative materials to be placed over thaw-susceptible soils, is scarce in many of the places where it is most needed, is costly to transport and the environmental effects of its removal and transport cause additional problems.[27]

There is a common tendency to predict the influence of major projects upon wildlife by estimating the amount of habitat that would be lost. For example, it has been pointed out that land portions of the sedimentary basins in Canada north of 60° total 470,000 square miles and that 1,350 square miles (0.3 per cent of the total) would be needed for seismic lines, roads, well sites, pipelines, campsites, storage dumps and airfields.[28] By simple arithmetic this would indicate that there is plenty of room left for wildlife habitat, but this conclusion must always be qualified with assurances that the actual area consumed is not critical migration or feeding or reproductive or staging habitat. For example, some species are crowded into small parts of the Arctic, such as the coastal zone of the Beaufort Sea where from mid-August until freeze-up practically the entire continental population of lesser snow geese is present.[29] Even small-scale damage to habitat could be critical in cases such as this. Providing certain critical wildlife and fishery habitats are avoided, there is a consensus that the greatest biological problem to be created by a pipeline would result from the improved access to areas along the route and increased hunting and fishing pressure near any new or expanded permanent settlements.

Combining the practices of our throw-away society with a landscape that is poorly designed for hiding debris and an environment with slow decomposition rates, we must identify waste disposal problems as another major environmental consideration for northern pipelines and all other northern development activities. Some of these expected environmental side-effects of northern petroleum development are discussed below in terms of controllability by extra expenditures.

C. ALTERATIONS EXPECTED TO BE EFFECTIVE FOR ENVIRONMENTAL PROTECTION

An appropriate question is whether the expenditure of, let us say, an extra one million dollars for the sake of lessening environmental impact will make any difference. A few examples are given here to indicate some of the steps that are believed to be effective for environmental protection. In only a few cases is it possible to indicate what these measures might cost. The next section of this paper will list some possible features of project deisgn that are less certain of being effective.

Hardy and Morrison[30] have shown that conventional pipeline construction practices may cause drainage problems and slope stability problems in northern areas with high ice content permafrost. If the problem conditions are recognized engineers may be able to avoid them by choosing better drained ridge locations and avoiding side-hill locations where the natural runoff pattern would be seriously disturbed. This avoidance technique, of course, is balanced against the economic objective of minimizing the length of pipeline. Hardy and Morrison do warn that the designer will never have sufficient data to be absolutely sure in all cases and there will be situations where choices involving very substantial amounts of money will have to be made, and in the final analysis unconventional and unproven solutions may have to be accepted. Using a slope stability example, these authors indicated that detailed designs for the approaches to river crossings where banks have high ice content permafrost have not been worked out for any of the proposed pipelines including the Alyeska line. In such cases the designer will need to consider more costly alternative modes of construction as compared to the conventional buried line. Possible solutions include construction on insulated surface pads, above-ground construction, the use of thermopiles to retain permafrost conditions, or ultimately the bridging of the whole valley. In other words, in those cases where minimal disruption to the environment is synonymous with minimal disruptions to the pipeline, technically and environmentally suitable solutions will be found, regardless of the cost. There is mounting evidence that there is corporate value in expenditures for a wide variety of environmental protection measures even when it involves environmental components that may not have a direct bearing on the integrity of the man-made structure itself. In other cases where there is forceful legislation, for example, with the Canada Fisheries Act which can require the construction of fish passage structures, similar environmental safeguards can be required even though their construction will have no specific benefit to the safety of the facility being constructed. Other examples could be the case of endangered species, such as falcons, or perhaps species for which there are international treaty commitments, such as waterfowl, and in these cases either public sentiment or international treaty obligations may result in expenditures for the sake of protecting the renewable resource in question without benefit to the facility or activity that threatens them.

Another control measure that will be effective in thaw-susceptible areas is the prevention of ponding of surface water. Because there are generally higher temperatures found under bodies of water as a result of the heat transport capability of the water, in the interests of terrain stability and in the interests of safety of the construction facility concerned, any extra expenditures that will prevent surface ponding of water, particularly over ice-rich perennially frozen ground, will be well invested. The design approach is to prevent ponding and thawing and to preserve the frozen ground condition.[25] Any expenditures to achieve these design objectives can be expected to result in long-term savings in maintenance.

In the case of roads, terrain damage can generally be avoided by building on five or more feet of gravel at a cost that may be as high as $200,000 per mile. However, not all terrain protection measures need be at great expense. For example, the revised practice of keeping bulldozer blades off the ground by attachment of suitable shoes at the corner of the blades is one that has greatly reduced surface disturbance and thermal erosion on winter-made seismic lines in the north without great additional expenditure. Another alteration that has not been prohibitively expensive but which has greatly reduced terrain disturbances is the voluntary shift to winter operations and the prohibition of operations on thawed terrain.

Another example of extra expenditure that has yielded definite returns in terms of environmental protection is the use of soft-wheeled vehicles in areas that have sensitive soils. The extra cost involved in developing new types of machines that result in less surface change to sensitive terrain may be returned to the investor because it will likely allow the start of fall work sooner and a later end to the work in spring than if tracked carriers alone were used.

For many of the physical disturbances in the north, the expenditure of more money through better engineering techniques will solve the problem. For example, the Institute of Arctic Environmental Engineering in Alaska has a project in which urethane foam is being tested as insulating material for roadbed construction at Prudhoe Bay, where a five to six foot gravel fill is normally required over exposed ground to prevent summer thaw. The urethane foam has approximately 30 to 50 times the design insulating value of gravel so that for five feet of gravel two inches of urethane foam can theoretically be substituted. In a 1,000 foot prototype road which has been in use since December 1969, there are indications that the urethane will perform successfully as a roadbed insulation material.[31] Although the cost of this design alternative is not documented for large-scale road construction projects, its effectiveness seems assured and in this case there would also be benefits to renewable resources such as fish or arctic foxes if the avoidance of gravel mining for road-building resulted in preservation of spawning beds or denning sites that might otherwise have been lost.

Features incorporated to protect either a pipeline or the surrounding environment will obviously be determined, in part, by what is acceptable to regulatory authorities. A case in point is whether or not regulatory authorities would accept as a warning system the time lag that is said to exist between the time of wrinkle and the time of fracture in a pipeline. For example, Hardy and Morrison[30] have indicated that the first development of a failure sequence is a buckling of the wall of the pipe and at this stress level there is a considerable factor of safety in relation to the stress system that would cause fracture. Trunk pipelines have been broken by washouts and landslides but experience has shown that a pipeline can withstand several feet of movement without becoming unserviceable. Tests on pipe proposed for the Alyeska pipeline indicated that the pipe will take 20 times as much deflection to develop a crack as it does to cause a wrinkle[32]. This suggests that there is a safety factor to give a warning system, assuming that there is a monitoring method to detect pipeline curvature or wrinkles, and that construction economies could be achieved by avoiding the need for expensive safeguards against pipeline fracture in places where they may never be needed. This question appears to be unresolved but it is one where large differences in cost could result from the level of safety stipulated.

A considerable degree of environmental protection can be achieved by appropriate timing of construction activities. The best examples are in the fields of fishery or wildlife protection where much can be achieved by avoidance of disturbance at biologically important stages. In a recently reported case from the village of Tuktoyaktuk, the local residents blamed their recent small whale catch on the suspicion that in the previous breeding season the whales were driven off by seismic test explosions in the area. In this case, the company responsible for the seismic exploration indicated that they could have timed the

exploration around the whale's breeding schedule, but the operators claimed that they did not know about these biological
time-scales[33].

When environmental protection requirements suggest a delay
in the timing of construction, these will obviously be extra costs
to the overall project. The magnitude of these costs is indicated
by estimates made by Canadian Arctic Gas Study Ltd.[34] on the
effects of a work stoppage at, for example, the mid-way point of
pipeline construction. The investment by that time would be
about 2.5 billion dollars and at 9 per cent interest the carrying
charge would be 225 million dollars per year. This means that
work stoppage at this point, in terms of interest on construction
capital alone, would cost more than 600,000 dollars per day.
Regardless of the extra cost, timing alterations may be an outright requirement by regulation or stipulation, but they may also
represent common sense. For example, in many areas of the
north it is better to build gravel roads in winter because construction operations will cause the least disturbance to tundra while it
is frozen. Surface preparation consists only of removing the snow
from the tundra surface before placing the gravel fill. It is also
possible to build gravel roads in the summer, provided there is
an access road to the gravel source, but gravel roads built in
summer will settle more and may require considerably more
gravel unless the tundra cover is preserved intact. In some cases
this may be impossible to achieve. During summer it is manda-
tory that all construction equipment be confined to the gravel
roadway, a condition that complicates and slows down road
building.[35]

One of the most effective ways to reduce terrain damage at
well sites or construction sites is to carefully select the site at a
time when the vegetation can be observed. Some of the documented cases[36] of terrain damage in the vicinity of well sites have
been where the rig site had been selected during the winter
without study of the surface conditions the previous summer.
The extra cost of a summertime trip for site selection, at a time
when it can be done more intelligently from an ecological point
of view, is a better investment than the cost of cleaning up or
trying to arrest thermal erosion that may result from poor siting
in the wintertime.

Another example of a protective measure that would be effective but which need not be prohibitively expensive is the regulation of low-altitude flight by aircraft, especially by jet helicopters,
in the vicinity of critical wildlife habitats. Both caribou and Dall

sheep are known to be frightened by the passage of jet helicopters,[37] as are migratory waterfowl such as geese and swans.[38] The establishment of minimal flight altitudes and restricted travel corridors in the vicinity of critical habitat will not be a great expense if these prohibitions are known well ahead of time; where such regulations result in extra mileage for helicopters, the cost can certainly be readily calculated.

Taking the estimate of a total cost of 1.5 billion dollars for the 1,550-mile Canadian section of a gas pipeline,[39] it appears that extra pipeline length required in the name of environmental protection will add a cost of approximately one million dollars per extra mile required. It is for this reason that the pipeline planners want as much advance notice as possible for areas that must be avoided by horizontal deviations in the route.

Another environmental protection cost is that associated with waste disposal at northern construction sites. One method of handling wastes at well sites in the north involves solid wastes from camp operations being burned regularly in open dumps at the edge of the gravel pad on which the camp is situated, and the residue is then covered with gravel in the spring without ever excavating a hole on the surface. For some well sites this covering of gravel has cost as high as $25,000.[40] For construction camps for 700 men it has been estimated that trenching requirements for burial of solid wastes would cost about $58,000 with an estimated cost of $2,000 for final covering and grassing. Operating costs would be about $112,000 for a total cost of $172,000 for one year of operation to prevent solid waste accumulation at such a site. By comparison, incineration for a camp of the same size would cost approximately $158,000, involving a $23,000 capital expenditure and $135,000 of operating expenses for a 52 week period.[40] These costs need not be for the sole purpose of meeting waste disposal standards. If there are proper garbage disposal mechanisms, wolves, grizzly bears, wolverine and foxes will be less vulnerable than they would be from untidy disposal operations. Insofar as an economic value can be placed upon these renewable wildlife resources, their protection through better waste disposal methods is a return on the expenditure for waste disposal.

As a final example, the phosphate esters that are used as fire-resistant lubricants in jet aircraft engines at compressor stations are suspected, because of their known toxic effects on humans and cattle, to be a potential threat to aquatic organisms if this material were to escape from a compressor station into streams. However, this hypothesis has not been tested for northern fish or aquatic invertebrates; even if there was a demonstration of a very significant biological threat from the inadvertent escape of these lubricants, this is an example of a case where an economic trade-off would probably enter into the final decision. The decision in such a case could involve the choice of industrial type turbines which, because they do not require the use of these synthetic lubricants, might be expected to be biologically safer, as opposed to the high temperature aircraft-type turbines which require the synthetic lubricants. Aircraft-type turbines have dominated the market in Canada for the last two years and the large units that are now available indicate that for a gas pipeline from Prudhoe Bay overland to the United States, involving approximately 75 turbines in a system using 1,500,000 horsepower, there would be a saving of approximately 4.7 million dollars per year over the cost that would result from use of the present generation of smaller-sized aircraft-type turbines that are currently used in southern Alberta.[41] In such a case any arguments about the environmental dangers of the synthetic lubricants that these power units require would have to be balanced against very significant cost savings. Furthermore, operators of compressor stations argue that the greatly decreased fire hazard that comes with use of synthetic fire-resistant lubricants far outweighs any biological hazards that may result from escape of these lubricants.

D. ALTERATIONS OF DOUBTFUL EFFECTIVENESS – THE POSSIBILITY OF UNAVOIDABLE ENVIRONMENTAL COSTS.

For northern pipeline construction, one of the areas of concern for which remedial practices are by no means assured is the question of extra pipeline safety required in multi-channel stream crossings. Secretary of the Interior, Rogers C. B. Morton, provided one of the most recent public pronouncements on this point when he argued that a Canadian route alternative for an oil pipeline from Alaska would be less desirable than one across Alaska because the Canadian alternative would involve "many more crossings of large rivers which, experience proves, are a major source of pipeline damage."[42] Secretary Morton's argument was based on the following premises: (1) that the main hazard during pipeline operation comes from floods which scour the river bed and bank and which, if large enough, may expose the pipe to buffeting from boulders and, thence, rupture; (2) that the wider the river, the greater the risk. It is possible to plan and design a pipeline crossing that would be safe for any particular stream by burying the pipe at adequate depth, by use of pipe of adequate thickness, or by adequate protective coatings on the pipe. However, a special circumstance in the north is that such a carefully designed approach does not necessarily take into consideration that in braided streams the streambeds are ever-changing; a well engineered stream crossing that ends up far removed from a new major channel of a watercourse is quite valueless. The engineering answer to this question is that stream stabilization plans will also accompany the major crossing sites. However, geomorphic and dendrochronological studies on extreme events in northern rivers[43] suggest that stream stabiliza-tion structures may be ineffective. Alberta recently lost a large mileage of railway to the floodplain of a river; in the north, too, extremely expensive engineering protective measures could quite easily be utterly ineffective, owing to the variable and extreme events that characterize the northern rivers that issue from moun-tainous areas.

There is no doubt that leaks from an oil pipeline would be a more serious environmental threat than leaks from a gas pipeline. But even for a gas pipeline, the cost of repairing line breaks, the fire hazard during gas escape, and the terrain damage that may result from repair operations all add to the importance of avoid-ing line breaks. The most elaborate and expensive preventative measures could be reduced in effectiveness when a pipeline facil-ity is located near to other man-made facilities such as roads. The list of equipment that is involved in gas pipeline damage is mainly road-associated equipment.[44] This means that if there is a public highway in the vicinity of a gas or an oil pipeline, the road will serve as one source of external disruption to any adjacent pipelines simply because many construction activities would use the road as their starting point. Governments are urging pipeline companies to reduce pipeline breaks but the largest single cause of these breaks is from outside forces: 25 percent of Alberta's 1971 oil pipeline breaks were from external causes; 20 percent of the 1971 oil pipeline breaks in the United States were from equipment rupturing the line; and, 68 percent of the 1971 gas pipeline breaks in the United States were from outside forces.[45] Those who argue that there would be environmental savings by routing of pipelines and roads in a common corridor must also take account of externally-caused pipeline accidents which would likely increase in frequency over what could be expected from a more isolated pipeline route.

There are other possible side-effects of multi-facility transportation corridors that could make more difficult any attempts to minimize environmental impact of a single pipeline. Cicchetti[17] in suggesting that there should be environmental savings if two or more pipelines were constructed at the same time in a common corridor overlooks the fact that it would be preferable, from a fisheries point of view, if no two crossings were made during the same year. This would permit at least partial recovery of the environment and fish populations before further disruption.[46] Similar questions arise in relation to possible interactions between pipelines, public highways, fires, and renewable resources that are influenced by fire. Although winter range of barren-ground caribou was damaged by fire before white man came to America, studies by Scotter[47] have indicated that the rate of destruction of range has increased with the growth of settlement and resource development; his studies indicate that fire reduces the winter range of barren-ground caribou but increases it for moose on upland forests. Whatever the net effect of fire on wildlife habitat, there will likely be design features to reduce the possibility of fires in the vicinity of pipelines. Very little direct study has been given to the effect of wildfires on pipelines. Fire is normally not viewed with much concern in the south where we now have pipelines, but in view of the fact that in some areas of the north up to 80 per cent of flow-slides are triggered by fires,[43] fires may be a threat to the stability of facilities such as a pipeline and the surrounding terrain. Such fire-induced flow-slides would probably become even more important to the stability of pipelines if a road were nearby because a road can be expected to increase the frequency of man-caused fires.

There is another feature of project design, this time involving vertical deviations in the depth of burying the pipe, that will likely have to be different in ice-rich perennially frozen ground than it is in southern Canada. In undulating terrain in the south it is common practice to scalp off the tops of hills to reduce the total length of pipe and to reduce the amount of required bending. The thermal erosion that results when an ice-rich knoll is exposed in the north will prevent this practice in many northern locations. For the safety of the pipeline alone, the southern practice of surface bulldozing will have to be avoided. Instead, there will apparently need to be acceptance of greater vertical deviations over undulating terrain, and therefore greater required length of pipe, or else greater depth of trenching over the high spots. Either alternative adds to the expense of construction. While either approach seems better than the risks that would result from the opening and levelling of ice-rich prominences, the effectiveness of these alternatives for avoiding serious terrain changes is not clear at this time.

A final example of alteration to project design that would be unpredictable in effectiveness is the speculation that the project could include caribou deflection mechanisms either to direct caribou to desired pipeline crossings or to direct them away from segments of a pipeline where caribou-induced eroison may pose a threat to the pipeline. Although the objectives of this suggestion are commendable, unpredictability of caribou, in terms of where they are going to be and when, would make it difficult to know where to build any such deflection structures or overpass or underpass structures. An incorrectly placed structure would obviously be of little value.

The examples given in this and the previous section suggest there are some features of the northern environment that prevent the advance design of fail-safe measures for all aspects of environmental protection. On balance, however, it appears that there are a large number of areas where environmental protection expenditures will be of definite benefit to renewable resources and to safety of man-made facilities.

E. SOME UNANSWERED QUESTIONS

Because the pipeline routes, and even many of the design features, depend upon sources and amounts of gas or oil and because these sources are not yet firmly established, there is not yet detailed information available for prospective northern pipelines. Therefore, there are bound to be questions that cannot even be perceived yet, let alone discussed; answers to specific questions must await the detailed studies that will accompany or follow proposals for precise location and design of northern pipelines.

Turning from prospective pipelines to broader aspects of northern resource management, there are several questions to be answered before there can be fully planned management of renewable resources. A knowledge gap very often identified by scientists is the lack of understanding about recovery times of northern ecosystems, how soon one event will follow another in natural systems, and how much one factor will change in response to a specified change in another factor.

In spite of the apparent emptiness of the north there is considerable evidence that those who choose to gain a livelihood from the land are living close to its productive capacity.[14] This is an important unanswered field of research for those who might have the task of defining what would be reasonable areas of land to guarantee that those so choosing have the option of a traditional way of life. It is evident that the allocation of so many acres per person based on experience in the south could be quite inappropriate in the north where it is known that the productive capacity of the land and biological growth rates are considerably less than they are further south on this continent. Viewing carrying capacity in terms of ability to absorb recreational visitors from the south, it is also evident that southern information on the carrying capacity of wildlands for recreation[48] is not directly applicable to the north without local documentation. To use one northern example, a dry lichen woodland would support very little foot traffic before destruction of the surface vegetation cover.

There is considerable attention given to the needs for restoration of fish and wildlife *habitat* after disturbances but there is less frequent mention of the need to restore wildlife *populations* in the event that they are locally exterminated[49]. As wildlife protection needs increase, management techniques will require attention to be directed to both habitats and populations. On the animal behavioral side, the knowledge that small piles of rocks and trees are known to deflect caribou into hunting traps[9] led to the hypothesis that man-made barriers such as an elevated pipeline might adversely affect caribou movements. This hypothesis has been countered by observations of caribou during post-calving loafing times in the Prudhoe Bay area where they did not appear to be markedly disturbed by man-made features, including simulated pipelines[50]. However, biologists both in government and in the pipeline consortia agree that to predict the possibility for deliberate deflection of caribou during migration time and in the vicinity of proposed pipeline routes still requires further knowledge of caribou responses at life-cycle stages other than just the post-calving period that was documented at Prudhoe Bay. It is also significant that the caribou studies, which have become the public focus of the environmental work in the north, have not involved estimation of the cost if one assumed that pipeline construction removed all of the caribou calf crop for one, two, three or some other number of years. Instead there has been a general assumption that *any* disruptions to caribou migrations and to caribou calving grounds are undesirable and should be avoided.

It is possible to predict many of the local environmental effects of a project such as a gas pipeline from the north; it is much more difficult to assess the environmental side-effects which might be triggered by the use of northern energy resources but which have a final impact on northern ecosystems only after passing through a complex chain of events in more industrialized parts of the continent. For example, it is known that airborne pollutants are particularly dangerous to non-rooted plants, such as lichens, because such plants receive their nutritional needs from airborne dust. Non-rooted plants increase in relative abundance as one progresses northward. This means that globally increasing levels of airborne contaminants will have a particularly important effect on northern vegetation – an ironic circumstance in view of the fact that we are entering a period when more of the energy that will produce these airborne pollutants will come from sources below those non-rooted arctic plants. The question of externally generated airborne pollutants is not part of the environmental studies that are to assess the side-effects of proposed northern development. Yet lead[51] and mercury[52] continue to increase in the north. There are two points to be made from these examples of global environmental changes. First, development of the fossil fuel reserves of the north will simply increase these changes in the background environment both through direct consumption of fossil fuel during northern activity and through use of these northern energy sources for southern industry which contributes to global atmospheric changes, including changes in the northern atmosphere. Second, the argument that a major asset of the north is its undisturbed state is now apparently true only in a relative sense because the value of the north as an area for scientific baseline studies has already been partly lost because of contamination by radioactive materials, lead and mercury. In summary, some of these global chemical changes which would be increased in magnitude or duration by consumption of northern energy resources will likely prove to be a greater environmental cost (because they are lasting and accumulative) than will the immediate and local disruptions of a project such as the pipeline because such a facility itself is a very short-lived object of perhaps only a few decades.

F. THE FUTURE FOR NORTHERN ENVIRONMENTAL MANAGEMENT

Costs to the public of environmental protection[53] and jurisdictional problems in international resource decisions involving oil and gas, water, migratory birds and animals, and marine resources[54] are deterrents to environmental mangement in the north. There have also been suggestions for changes in administrative and operational procedures to bring about better environmental management. For example, aesthetic disruption to tundra would be reduced if there were ways of reducing the intensity or repetitiveness of seismic exploration. Similarly more widespread use of study consortia (as opposed to larger numbers of smaller operators, each with their own airstrips, roads and camps) and changes in the administrative work requirements for exploration permits could also reduce the overall demand placed upon northern areas by exploration or technical study groups.

From an ecological viewpoint, which is by definition longterm, one could argue that the hurried approach to the search for fossil fuels is a blessing in disguise because it will mean that northern hydrocarbon extraction will be a short-lived event of perhaps only two or three decades in what is otherwise a longterm, post-glacial or inter-glacial evolution of northern ecosystems. A short, intense production period which is so important for maximization of economic benefits, may in the end be the best for the northern environment. Perhaps we should be saying, 'If we must have this frontier frenzy, let it be over with and out of the way as quickly as possible.' As tempting as this argument may be, one must also consider the need to conserve the petroleum resources of the north for use by future generations in ways not yet imagined.

Whatever pace is adopted for removal of the non-renewable northern fuel resources, the need will remain for long-term management of the renewable resources. Current government policy centres around the concept of 'managed use'[55]. Here the basic premise is utilization but under conditions which minimize alteration of the resource-base while accepting the fact that if the land is to be used, some degree of disturbance is unavoidable. One thing seems certain amongst environmental scientists and that is the realization that we simply must understand how natural systems function.[56] Ives[57] has aptly described the current situation, whereby man, the scientist, has made tremendous progress in learning to understand the very large, the universe, and the very small, the molecule, through application of telescope and microscope. What is missing is a "mesoscope" that would allow us to better understand the environment with which we directly interact and upon which our very existence depends. From there, the goal would be to bridge the gaps between technically sound information, various management techniques, and the selection of desirable alternatives of use for renewable and non-renewable resources.

FOOTNOTES

1. F. Kreith, 1973. Lack of impact. Environment 15 (1): 26-33. (Note that this poor record of effectiveness for environmental impact statement exists despite the fact that the United States, unlike Canada, has a statutory requirement – through Sec. 102 of the U.S. National Environmental Policy Act – for consideration of "presently unquantified environmental amenities and values in decision-making along with economic and technical considerations").
2. D. M. Kiefer, "Assessing technology assessment," *The Futurist* (December, 1971) 234-239.
3. D. A. Chant, "Our costly cleanup," *Financial Post* (3 March, 1973) 13.
4. J. R. Mackay, "Disturbances to the tundra and forest tundra environment of the western arctic," *Can. Geotech. Jour.,* Vol. 7 (1970), 420-432.
5. G. Feinberg, "Long-range goals and the environment," *The Futurist,* (December, 1971) 241-246.
6. R. M. Strang, "Studies of vegetation, landform and permafrost in the Mackenzie Valley. The effects of disturbances; some case histories." Manuscript, Northern Forest Research Centre, Environment Canada (Edmonton, 1973), 37 p.
7. For a fuller discussion of this conclusion see R. O. Brinkhurst, and D. A. Chant. *This good, good, earth: Our fight for survival* (Toronto: Macmillan Co. of Canada, 1972) 174 p.
8. For an excellent review article, see N. H. Coomber, and A. K. Biswas. *Evaluation of environmental intangibles* (Bronxville, N.Y., 1973), 77 p. For a more specific example of this problem, see P. H. Pearse. "Some economic and social implications of the proposed Arctic International Wildlife Range," *Univ. of B.C. Law Review,* Vol. 6, No. 1, Supplement (1971), 36-46.
9. R. M. Hill, "The arctic environment and petroleum pipelines," *The Musk-Ox,* Vol. 9 (1971), 35-41.

10. A. C. Fisher, J. V. Krutilla and C. J. Cicchetti, "The economics of environmental preservation: a theoretical and empirical analysis," *American Economic Review,* Vol. 62, No. 4 (1972), 605-619.

11. P. H. Pearse, "A new approach to the evaluation of non-priced recreation resources," *Land Economics,* Vol. 44, No. 1 (1968), 87-99.

12. Environment Protection Board, "What is the value of wilderness?," *Environment Protection Board Gas Pipeline Newsletter* (October 1, 1972).

13. W. N. Irving, and C. Harington, "Upper Pleistocene radiocarbon dated artefacts from the northern Yukon," *Science,* Vol. 179 (1973), 335-40.

14. J. K. Naysmith, *Canada North – man and the land* (Ottawa: Department of Indian Affairs and Northern Development, 1971), 44 p.

15. It should be added parenthetically that Naysmith is aware of the counter-arguments to this approach, in that revenue that would accrue to the nation from geophysical exploration and pipeline development and maintenance has not been considered in the calculations. It is certainly true that sales tax from the pipe and corporation taxes from operation of the pipeline would be a factor of national importance. However, such arguments mean little to the actual community concerned and Naysmith has very appropriately defended his right to make the comparison on the basis of the specific community concerned.

16. See, for example, Travacon Research Limited, *Economic study of transportation in the Mackenzie River Valley,* Vol. 1 (Calgary: Travacon Research Limited, 1972), 281 p.

17. C. J. Cicchetti, *Alaskan Oil: Alternative routes and markets* (Baltimore: The Johns Hopkins University Press, 1972), 142 p.

18. J. E. Maybin, "Northern energy resources," in, *Proceedings, Fourth Northern Resources Conference* (Whitehorse, Yukon, 1972), pp. 23-25.

19. An example of a study that had to rely heavily on estimates of probability is that by Quirin and Wolff which estimated the probability of an oil spill for each transportation alternative, as well as the magnitude of such spills and the cost of cleanup, and thus calculated approximate environmental costs for each alternative. The environmental costs thus calculated were 9 cents per barrel for tankers using the Northwest Passage, 1 cent for the Trans-Alaska pipeline and for one alternative using aircraft, and less than 1/10 of one cent per barrel for other alternatives (including the Mackenzie Valley pipeline). The pollution costs which Quirin and Wolff estimated were small relative to the magnitude of transportation cost estimates, and were not large enough to reverse decisions made on purely economic grounds.

(G. E. Quirin, and R. N. Wolff, "The economics of oil transportation in the Arctic," Working paper (1971), 71-16, prepared for a conference on Canadian-U.S. Law of the Sea Problems, June 1971, University of Toronto, School of Business. Original not seen, quoted from U.S. Department of the Interior, *An analysis of the economic and security aspects of the Trans-Alaska Pipeline,* Vol. 1, Summary (Washington D.C., Office of Economic Analysis, 1972).

20. G. Shaw, Alyeska pipeline, p. 82-86. IN: Proceedings, Fourth Northern Resources Conference (Whitehorse, Yukon, 1972), pp. 82-86.

21. D. P. McDonald, "Oil and gas pipelines," in, *Proceedings, Fourth Northern Resources Conference* (Whitehorse, Yukon, 1972), pp. 71-74.

22. J. E. Bryan, C. E. Walker, R. E. Kendel and M. S. Elson, "The influence of pipeline development on freshwater aquatic ecology in northern Yukon Territory," Progress report on research conducted in 1971 (Vancouver. Manuscript, Fisheries Service, Environment Canada, 1972,) 45 p.

23. See, for example: 1) W. A. Fuller, and P. G. Kevan, Eds. "Productivity and conservation in northern circumpolar lands," I.U.C.N., New Series No. 16. (Morges, Switzerland, 1970), 344 p.; 2) A. H. Macpherson, "Northern ecology and development," Manuscript for Paper presented at Shell Program on the Canadian North (University of Toronto, 30 January, 1971), 31 p; 3) M. J. A. Butler, and F. Berkes, "Biological aspects of oil pollution in the marine environment. A review," Manuscript Report No. 22, Marine Sciences Centre, McGill Univ., Montreal (1972) 118 p; 4) R. F. Leggett, and I. C. MacFarlane, Proceedings of the Canadian Northern Pipeline Research Conference, 2-4 February 1972. N.R.C. Tech. Memorandum 104 (Ottawa, 1972), 331 p; 5) P. Roberts-Pichette, Annotated bibliography of permafrost-vegetation-wildlife-landform relationships. Information Report FMR-X-43. (Ottawa: Forest Management Institute, Environment Canada, 1972), 350 p; 6) L. C. Bliss, and E. B. Peterson, "The ecological impact of northern petroleum development," Proceedings of Fifth International Congress, Arctic Oil and Gas: Problems and Possibilities, Fondation Francaise d'Etudes Nordiques, LeHavre, France. (In press); 7) D. H. Pimlott, K. M. Vincent and C. E. McKnight, "Arctic Alternatives. Proceedings of a National Workshop on People, Resources and the Environment North of 60," Canadian Arctic Resources Committee (Ottawa, 1973), 391 p; 8) In addition, the Organization for Economic Co-operation and Development has recently compiled a check list of environmental problems associated with production of primary energy sources, with use of fuels in dispersed installations, and with use of fuels for generation of electricity (O.E.C.D., "Environ-

mental impacts arising from energy production and use," Unpublished File 73(9) of o.e.c.d. Environmental Committee (Paris, 1973).

24. Roberts-Pichette, *op. cit.* (See fn. 23 above).
25. C. B. Crawford, and G. H. Johnston, "Construction on permafrost,"*Can. Geotech. Jour.*, Vol. 8, No. 2 (1971) pp. 236-251.
26. T. G. Watmore, "Thermal erosion problems in pipelining," in, Proceedings of Third Conference on Permafrost, Associate Committee on Geotechnical Research, Technical Memorandum 96, (Ottawa: National Research Council, 1969), pp. 142-162.
27. A. Feingold. "Problems of pipeline construction under arctic conditions," Paper presented at 74th National meeting of American Institute of Chemical Engineers, New Orleans (11-15 March 1973), 8 p.
28. G. Rempel, "Arctic terrain and oil field development," in, W. A. Fuller and P. G. Kevan, eds., "Productivity and Conservation in Northern Circumpolar Lands," i.u.c.n., New Series, No. 16 (Morges, Switzerland, 1970), pp. 243-251.
29. Special Habitat Evaluation Group, "An inventory of wildlife habitat of the Mackenzie Valley and the northern Yukon for the Environmental-Social Program, Northern Pipelines," Manuscript (Edmonton: Canadian Wildlife Service, Environment Canada, 1973), 484 p.
30. R. M. Hardy, and H. L. Morrison, "Slope stability and drainage conditions for Arctic pipelines," in, *Proceedings of the Canadian Northern Pipeline Research Conference.* (Ottawa: N.R.C. Technical Memorandum 104, 1972), pp 235-248.
31. Institute of Arctic Environmental Engineering, *Annual Report,* (Univ. of Alaska, College, Alaska, 1971) 16 p.
32. J. P. D'Donnell, "Alyeska line commands careful design," *Oil and Gas Journal* (November 29, 1971), p. 24-28.
33. W. Kornberg, "Concern for the artic environment," *Science News,* Vol. 97 (1970) pp. 486-488.
34. From a presentation by V. L. Horte, President, Canadian Arctic Gas Study Ltd. to Canadian Pipeline Contractors Association, Vancouver, B.C., 12 April 1973. Although these estimates of the effect of a work stoppage were made in the context of delays resulting from labour problems, it is unlikely that delays resulting from environmental protection requirements would change the arithmetic of the estimates.
35. W. P. Stokes, "North Slope – construction criteria for roads and facilities," *Jour. of Petroleum Technology* (October, 1971), 1209-1217.
36. J. D. H. Lambert, "Botanical changes resulting from seismic and drilling operations, Mackenzie Delta area," *ALUR* 71-72-12. (Ottawa: Department of Indian and Northern Affairs, 1972), 70 p.
37. A. W. F. Banfield, "Anticipating the effects of a buried gas pipeline on the northern ecosystem. *Science Forum* 25 (1972), pp 15-17.

38. T. W. Barry, and R. Spencer, Wildlife response to oil well drilling. Unpublished report, Can. Wilflife Serv., Edmonton (1972), 39 p.
39. J. C. Osler, "Arctic pipelines," in, *Proceedings, Fourth Northern Resources Conference* (Whitehorse, Yukon, 1972), pp. 87-89.
40. A. N. Gunter, "The management of solid waste disposal in the northern environment as it relates to resource development," Unpublished paper presented at the Engineering Institute of Canada annual meeting (October, 1972), 35 p.
41. R. C. Stauffer, "Gas turbine fills the market to date," *Canadian Petroleum,* (September, 1971), pp. 43-45.
42. See the News Release, 5 April 1973, which took the form of a letter from the office of the Secretary, United States Department of the Interior, to all Members of Congress.
43. D. K. MacKay, "Summary technical report on hydrologic aspects of northern pipeline development," (Ottawa: Glaciology Division, Environment Canada) 13 p. (In press, Environmental-Social Program, Task Force on Northern Oil Development).
44. Gas Arctic Systems Study Group, "Transportation corridor study," Final Report. Physical descriptions – Technical considerations. Vol. II. Prepared by Pemcan Services, Calgary, 1971.
45. Alberta's Environment Minister recently stated (*Edmonton Journal*, p. 51, 1 March 1973) that of the province's 83 pipeline breaks in 1971, outside causes resulted in 21 spills, external pipeline corrosion caused 20, installation failures caused 15, pipe failure from pressure of pumping caused 13, internal corrosion caused 7, and miscellaneous events caused 7 breaks. Recent statistics for causes of 1023 pipeline breaks in the U.S. in 1971 are given in the Editorial, *Oil and Gas Journal* (April 10, 1972).
46. J. N. Stein, C. S. Jessop, T. R. Porter and K. T. J. Chang-Kue. "An evaluation of the fish resources of the Mackenzie River Valley as related to pipeline development," (Winnipeg: Fisheries Service, Environment Canada, 1973), 113 p. (In press, Environmental-Social Program, Task Force on Northern Oil Development).
47. `G. W. Scotter, "Fire, vegetation, soil, and barren-ground caribou relations in northern Canada," in, *Proceedings – Fire in the northern environment – a symposium* (Portaland, Oregon: Forest Service, U.S. Dept. of Agriculture, 1971), pp. 209-230.
48. J. A. Wagar, "The carrying capacity of wildlands for recreation," Forest Science Monograph No. 7. (1964), 24 p.
49. Yukon game officials have pointed out that 75 years ago sheep populations were locally eliminated in the Dawson area. After such a long period they have not yet returned even though the necessary habitat is there, unmodified but "empty."
50. K. N. Child, A study of the reaction of caribou to various types of simulated pipelines at Prudhoe Bay. Alaska. Manuscript of Paper presented at First Int. Symposium on Ungulate Behaviour and its Relation to Management (Calgary, 1972), 15 p.
51. C. S. Cook, "Energy: planning for the future," *American Scien-*

tist Vol. 61, No. 1 (1973), pp. 61-5.

52. H. V. Weiss, M. Koide and E. D. Goldberg, "Mercury in a Greenland ice sheet: evidence of recent input by man," *Science*, Vol. 174 (1971), pp. 692-4.

53. For an example of the additional costs expected to be incurred by the State of Alaska in habitat protection and enforcement of fish and game laws see A. R. Tussing, G. W. Rogers, V. Vischer, R. Norgaard and G. Erickson, "Alaska pipeline report. Alaska's economy, oil and gas industry development, and the economic impact of building and operating the Trans-Alaska pipeline," Manuscript Copy of Report 31, Institute of Social, Economic and Government Research, University of Alaska, (1971), 123 p.

54. A. R. Thompson, and H. Eddy, "Jurisdictional problems in natural resource management in Canada," Special Study, Essays on Aspects of Resource Policy. Science Council of Canada, Ottawa. (Quoted from Science Council of Canada. 1973. Natural resource policy issues in Canada. Report No. 19. 59 p.), In Press.

55. J. K. Naysmith, "Management of Polar Lands," Unpublished paper presented at 12th Technical Meeting of International Union for Conservation of Nature and Natural Resources, Banff, Alberta (1972), 21 p.

56. F. K. Hare, "The natural environment of the Canadian North," Unpublished Report, (Ottawa: Environment Canada, 1972), 56 p.

57. J. D. Ives, "Arctic tundra: How fragile? A geomorphologist's point of view," Trans. of the Royal Society of Canada, Series IV, Vol. 8 (1970) pp. 401-404.

8.
IMPACT OF A MACKENZIE PIPELINE ON THE NATIONAL ECONOMY

John Helliwell

This chapter presents a quantitative assessment of the impact on the national economy of a Mackenzie Valley gas pipeline and associated production of natural gas. The research is the result of the joint effort of many persons,[1] and is closely related to the cost-benefit analysis reported in Chapter 10. Both chapters use simulation results from an expanded version of the RDX2 quarterly model of the Canadian economy,[2] but focus on rather different issues. The simulations in the present chapter attempt to picture the difficulties posed for the economy in digesting a large-scale investment of the sort being proposed by the Arctic Gas consortium, as described in Chapter 2. The results shed some light on the possible short term consequences for income, employment, prices, the exchange rate, and the various components of Canada's foreign balance of payments. By contrast, in Chapter 10, we by-pass these immediate effects on the national economy, and study the longer term costs and benefits, using simulation of the pipeline and gas production over a time horizon exceeding thirty years.

Both types of study are important if the pipeline question is to be treated seriously. No large project ought to be approved unless it can be accomplished without causing excessive distortions and adjustment problems for the national economy. Furthermore, no project that makes use of non-renewable natural resources should be approved unless it promises to be the best use of those resources, taking due account of future as well as present benefits, costs, and opportunities. The present chapter deals only with the former – the problems of macroeconomic adjustment: these logically come first because the results on this score provide some of the necessary material for the broader evaluation of costs and benefits in Chapter 10.

This chapter contains four main sections. Section A describes in quantitative detail the construction, financing, and operation of a gas pipeline matching closely the Gas Arctic Consortium's proposal described in Chapter 2. This section also contains the modelling of, and a commentary on, the National Energy Board (NEB) procedures used in setting pipeline tariffs. The calculations run from 1973 until 2011, the assumed end of the pipeline's useful life. Section B deals with the production of gas in the Mackenzie Delta, on the assumption that Canadian gas is used to fill that half of the pipeline's capacity not used to trans-ship Prudhoe Bay gas. Section C deals with the macroeconomic effects of the investment and operating phases of the pipeline and gas production sectors described in sections A and B. Section D contains a summary of, and qualifications to, the main results.

A. MODELLING THE PIPELINE

There are two main aspects to our modelling of the construction and operation of a Mackenzie Valley pipeline. First, it was necessary to design a set of internally consistent equations describing the main features of the pipeline. We can then start with certain primary (or "exogenous") information about the timing and costs of construction and use the equations to determine the financing, pipeline tariffs, direct balance of payments effects, and so on. If the set of equations is self-contained, as ours is, it can be run forward through the entire life of the pipeline, producing results of the sort included in Table 1.

The second aspect of pipeline modelling is the linkage between the pipeline and the rest of the economy. In our research, this linkage is established by altering a number of equations of a large general, or macroeconomic, model of the Canadian economy (known as RDX2) to include the relevant aspects of pipeline expenditure, employment, financing and operation.[2] These links, which are described in Appendix C, enable the national model to be run with and without the pipeline sector to show the macroeconomic repercussions of the pipeline. Section C of this chapter contains the results of such simulations for the pipeline and Mackenzie Delta gas production treated together.

The pipeline sector is depicted by a set of 25 equations showing the assets, liabilities, financing, and rate determination for about 2400 miles of pipeline from the Alaskan border and from the Mackenzie Delta, joining near Fort Simpson, to deliver gas to the United States border at two points, as shown in the map in Chapter 2. At full capacity, the pipeline is assumed to take in 4.5 billion cubic feet per day (henceforth bcf/d); to use .5 bcf/d for compression and refrigeration; and hence to deliver 4.0 bcf/d to the southern ends of the pipeline. The total cost of the project, including interest until the end of 1978, amounts to more than $5.3 billion measured as the sum of dollar amounts expended, at then current prices, during the construction period. The financing is assumed to involve a 51/49 split between Canadian and American equity ownership, a 4/1 ratio of debt to equity, and the use of foreign sources for about two-thirds of the debt financing. The construction period is assumed to run from mid-1975 until the end of 1978, with the initial gas flow from the Mackenzie Delta starting in the second quarter of 1978.[3]

The many assumptions that underlie our calculation of the total cost of the pipeline are described in detail in Appendix A.

On the basis of our cost and timing assumptions, we constructed equations showing, among other things, financing requirements (FINREQ), the rate-base value of the pipeline (KPLBASE), the total amount of pipeline revenue (YPT$), corporation taxes paid by the pipeline (TCP), the pipeline tariff rate (TARIFF), and the price of gas delivered to the United States border (PGASDEL). The tariff rate and the delivered price of gas are both measured in Canadian cents per thousand cubic feet (¢/mcf). The tariff rate and total tariff revenues do not include the cost of gas used in transmission. The delivered cost of gas includes the wellhead cost of all input gas, taking due account of the amount used in transmission. Wellhead prices are based on existing contracts, as explained in Appendix B. The annual values of the six series, from 1973 to 2011 inclusive, are shown in Table 1. The underlying model is quarterly; the annual values are sums

TABLE 1
Mackenzie Valley Pipeline—Selected Simulation Results
(Annual Values in Current Dollars)

Year	Financial Requirements ($ mill.) FINREQ	Pipeline Rate Base ($ mill.) KPLBASE	Pipeline Tariffs ($ mill.) YPT$	Corporation Tax Payable by Pipeline ($ mill.) TCP	Pipeline Tariff ($/mcf) TARIFF	Price of Gas Delivered to South ends of the pipe line (¢/mcf) PGASDEL
1973	27	27	0	0	0	0
1974	29	56	0	0	0	0
1975	1219	1276	0	0	0	0
1976	1521	2798	0	0	0	0
1977	1327	4125	0	0	0	0
1978	1213	5369	75	0	53	89
1979	121	5208	332	0	50	86
1980	-168	5047	617	0	49	87
1981	-292	4886	720	0	49	87
1982	-312	4725	716	0	49	87
1983	-161	4564	708	0	48	86
1984	-161	4403	697	0	47	85
1985	-161	4242	686	0	46	90
1986	-161	4080	676	0	46	89

Year						
1989	-161	3597	701	55	53	97
1990	-161	3436	784	149	53	103
1991	-161	3275	782	156	53	103
1992	-161	3114	780	163	53	102
1993	-161	2953	778	170	53	102
1994	-161	2792	775	176	53	102
1995	-161	2631	773	182	52	108
1996	-161	2470	770	187	52	107
1997	-161	2308	767	192	52	107
1998	-161	2147	765	197	52	107
1999	-161	1986	762	202	52	107
2000	-161	1825	760	206	52	109
2001	-161	1664	758	210	51	111
2002	-161	1503	756	214	51	113
2003	-161	1342	754	217	51	116
2004	-161	1181	753	220	51	118
2005	-161	1020	746	220	50	120
2006	-161	859	720	205	48	121
2007	-161	698	692	189	46	122
2008	-161	536	665	172	44	123
2009	-161	375	639	155	43	124
2010	-161	214	613	138	41	126
2011	-161	53	587	121	39	127

of quarterly flows (such as revenues or taxes) averages of quarterly prices (such as the price of gas) and the end-of-year values for stocks (such as the rate-base value of the pipeline).[4]

Some explanation may be required for the pipeline tariff, which has some unusual movements. Our equation for TARIFF attempts to duplicate the accounting used by the NEB in making its decisions about pipeline tariffs. The tariff is set high enough to cover, on a year-by-year basis, straight-line depreciation of the rate base at the end of 1978, direct operating expenses, interest on debt, corporation income taxes, and a 14 per cent return on the book value of equity. The allowance for corporation income taxes, which causes a tariff jump in 1989, is assumed by pipeline operators, but has not been formally accepted by the NEB, because no major gas carrier has been paying taxes at the time of a rate hearing.[5]

In the absence of changes in the debt/equity ratio, the NEB procedure would result in a tariff rate continually dropping,[6] in current cents per mcf, for a service whose real value is presumably constant and whose money value should correspondingly be rising. In our calculations, the full effects of this illogical pricing procedure are obscured by the assumed financing pattern. Under our assumptions, cash flow in excess of current costs is used to retire debt, and no equity is retired until 2005, by which time all of the debt has been paid off. The declining debt/equity ratio has the effect of raising the tariff rate above what it otherwise would be, because the cost of equity capital before corporation taxes is assumed to be over 25 per cent, while the average cost of debt is about 9 per cent.

That much substitution of equity for debt also seems illogical, as the relatively high after-tax rate of return on equity is presumably justified only by a substantial use of debt to increase the riskiness of the equity return. If the NEB did not permit the lowering debt-equity ratio, the tariff would drop faster than we have indicated, as would the forecasts of corporation taxes. If the tariff were set as a price per unit of throughput, fixed in constant dollars, then the initial tariff in ¢/mcf would be lower at the beginning than at the end, for any given debt/equity ratio. We can easily alter our present calculations if any changes are thought likely to be made in the NEB tariff-setting procedure.

One reaction to earlier versions of this paper has been that the tariff-setting procedures could not be as illogical as suggested above, because pipeline tariffs have in the past remained constant or grown in nominal terms. Thus, it is argued, our analysis has not been applicable. However, the illogic of the NEB tariff-setting procedure has been obscured because pipelines have been continually expanding; and the new additions to the rate-base, coming in at higher nominal cost, have prevented the average tariff from falling. The fact that the tariff is always higher, in real terms, for the early users than for the later users of the pipeline (and higher on average than it ought to be) will only show up in the average tariff when pipelines cease to expand, as inevitably they must. We plan subsequent studies showing how the pipeline tariffs would look if they were set as a constant real charge per mcf transported, taking due account of the effects of terrain and latitude on construction costs.

In part because of the tariff-setting procedures described above, the delivered price of gas (PGASDEL) does not rise at a rate as high as the 4 per cent assumed rate of increase of prices and costs in general. PGASDEL is an estimate based on established contracts for wellhead prices, and on "delivery" of the gas only as far as the southeast corner of Alberta. The corresponding cost of gas delivered to Toronto city gates in 1980 would be about 30¢/mcf higher than PGASDEL.

B. GAS PRODUCTION IN THE MACKENZIE DELTA

The extent to which Mackenzie Delta gas would be transported in the pipeline described in Section A depends on, among other things, the amount and desired production rate of natural gas associated with the Prudhoe Bay oil reserves. As is well known, the timing and transport routing for Alaskan oil development are still uncertain. The Gas Arctic news releases have consistently stated that 50 per cent of the pipeline throughput would come from Prudhoe Bay and the other half from the Mackenzie Delta. As far as we can determine, this 50/50 split could not be achieved during the early years of pipeline operation without using some of the Prudhoe Bay "cap gas" usually left until the bulk of the oil has been extracted. In our earlier studies, we assumed that the Alaskan conservation authorities would not permit the early withdrawal of cap gas, and hence that Mackenzie Delta gas would be used to provide substantially more than 50 per cent of the pipeline's throughput in the early years of its proposed operation. We have since been advised that it may be feasible (if it proves acceptable to the conservation authorities) to withdraw cap gas in the early stages, replacing it with salt water injected to maintain pressure. Thus earlier, if more expensive, extraction of larger volumes of Prudhoe Bay gas could be accomplished. The simulations reported in this chapter therefore assume that Prudhoe Bay and the Mackenzie Delta each provide 50 per cent of the pipeline throughput from the outset to the completion.

Some of the main features[7] of the gas production operations are summarized in Table 2. All of the figures are annual flows in millions of current dollars. YGAS\$ is the total revenue from wellhead sales of Delta gas. TROYAL and TCGAS measure royalties and corporation taxes, respectively, received by the federal government. YGASB represents accounting profits of the gas producers, after deducting royalties, taxes, and interest payments. These amounts are of interest in themselves, but they do not represent the net benefits of the project to any of the participants, as they do not take account of what the productive factors would receive if, in the absence of this project, they were employed elsewhere. This broader question will be dealt with in Chapter 10.

Of special interest are the last two series in Table 2. The series, which are defined in Appendix c, provide a handy summary of the direct balance of payments impacts of the pipeline and related gas production. XBALP&G\$ shows the direct trade account impact of the combined pipeline and gas production sectors, taking account of export receipts from gas sales and pipeline tariffs, net of interest and dividend payments to foreign holders of pipeline securities, and the direct import content of pipeline and gas investment. The series shown in Table 2 are based on the assumption that about half (13 tcf) of the Mackenzie Delta gas transported would be exported, with the rest used in Canada. This is consistent with the current Arctic Gas view, as presented in Chapter 2, that 10 to 15 tcf of Mackenzie Delta gas would be exported.

The direct capital account impact of the combined operations is shown as FBALP&G. The positive elements of this series include foreign sales of pipeline bonds and shares and the foreign interest-free loans negotiated by gas producing firms. The negative elements comprise repayment of loans and retirements of foreign-held securities. The sum of the two series shows the net balance of payments impacts of the investment and operating phases of pipeline operation and gas production. For example, in 1985, when gas production, debt retirement, etc., are in full swing, the net balance of payments impact is about $400 million per year. This implies almost as big a positive influence on the demand for Canadian dollars as at any time during the construction period, when the direct import requirements and the capital inflows are largely offsetting. If we were to make use of higher (and more realistic) export gas prices than those contained in present contracts (as described in Appendix B), or if more of the Delta gas were to be exported, the net balance of payments impact during the operations phase would be even greater. For example, if the 1985 wellhead price were 48 ¢/mcf, rising thereafter at the general rate of price increase, and if all of the Delta gas were exported, the balance of payments surplus would be almost $700 million, rising to more than $1.2 billion by the year 2000.

The sum of XBALP&G$ and FBALP&G provides a simple measure of the required amount of reduction in other net exports and capital inflows. Under a flexible exchange rate, a change in the exchange rate is the primary mechanism whereby these accommodating trade and capital flows would be induced. The size of the required change in the exchange rate is a frequently used measure of the balance of payments effects of the pipeline and gas exports. Although the likely size of such changes will be discussed in the next section, most readers will find it easier to interpret the basic flow data of Table 2, rather than to focus on the size of the resulting exchange rate change. The important thing to remember is that in equilibrium the exchange rate has to move enough to effect offsetting changes elsewhere in the balance of payments. We turn now to consider the simulation results showing the pipeline and gas production in the context of the RDX2 model of the Canadian economy.

C. MACROECONOMIC REPERCUSSIONS

The first point to be made is that the simulation of an aggregate econometric model does not provide any measure of the costs and benefits of a large project. For one thing, it is difficult to translate movements of aggregate employment, incomes, prices and wages into any coherent measure of national advantage. More fundamentally, if the economic effects of a project can be foretold, then monetary and fiscal policies can be designed to offset, in the context of the model, any troubling consequences.

TABLE 2
Mackenzie Delta Gas Production—Selected Features
(Annual Values in Millions of Current Dollars)

Year	Sales of Processed Gas	Royalty Payments	Corporation Tax Payable On Gas Production	Book Profits Gas Production After Royalties And Taxes	Balance of Payments Effects of Pipeline and Gas Production	
					Direct Capital Account Impacts (Surplus if +ive)	Direct Trade Account Impacts (Inflow if +ive)
	YGAS$	TROYAL	TCGAS	YGASB	XBALP&G$	FBALP&G
1973	0	0	-5	-19	-14	26
1974	0	0	-5	-19	-14	30
1975	0	0	-3	-5	-623	597
1976	0	0	-9	-9	-641	927
1977	0	0	-91	-15	-507	905
1978	26	1	-97	-11	-251	857
1979	118	5	-13	22	136	329
1980	237	11	18	75	457	-101
1981	279	13	28	89	585	-222
1982	279	13	29	84	596	-235
1983	279	24	24	68	528	-136
1984	279	27	26	67	527	-136
1985	320	32	41	98	566	-136

1987	320	32	37	85	563	-136
1988	320	32	35	79	561	-136
1989	320	32	33	72	602	-136
1990	361	36	47	102	712	-128
1991	361	36	44	95	716	-104
1992	361	36	42	89	720	-104
1993	361	36	40	82	724	-104
1994	361	36	37	75	728	-104
1995	402	40	50	104	773	-104
1996	402	40	47	97	777	-104
1997	402	40	44	88	780	-104
1998	402	40	40	80	784	-104
1999	402	40	37	71	787	-104
2000	418	41	39	76	807	-104
2001	435	43	42	82	827	-104
2002	452	45	45	87	848	-104
2003	470	47	47	93	869	-104
2004	489	48	50	99	892	-104
2005	509	50	53	105	912	-91
2006	529	52	56	111	919	-78
2007	550	55	59	118	925	-78
2008	572	57	62	125	933	-78
2009	595	59	65	132	942	-78
2010	619	61	68	139	952	-78
2011	644	64	72	147	963	-78

But if any shock to the aggregate economic system caused by a large project can be offset by monetary and fiscal policies, it could also be duplicated by such policies. If so, then how could any aggregate model show net benefits from a large project? The short answer is that it could not. To assess net benefits, we must use procedures like those employed in the final chapter of the book.

Then why is it worthwhile to simulate the macroeconomic consequences at all? For two reasons, I think. On the one hand it enables policy-makers, business planners, and citizens alike to see the extent and timing of the transfer of real resources into and out of investment in the project, and allows the design of sensible policies to help the economy digest large projects. Secondly, the magnitude and nature of the macroeconomic effects may give some clues as to where serious bottlenecks and regional disparities might arise.

Thus forearmed, the reader is invited to consider the results of some of our simulation experiments. To help reveal the indirect and direct consequences for the rest of the economy, we have used simulations from 1973 to 1985 of the aggregate quarterly model RDX2. Appendix C explains the assumptions and procedures, while Tables 3 and 4 show the macroeconomic effects of the pipeline and gas production, as cushioned by selected monetary and fiscal policies. Our method was to first run the model from 1973 to 1985 without the pipeline and gas developments. This is called the 'control simulation', because it provides a basis for comparison. Next we ran the model again with the "PIPE & GAS" sector in operation, with full linkages between that sector and the rest of the economy. This is called the 'shock simulation' because it pictures the economy under the influence of a disturbance in the form of the northern gas developments. Tables 3 and 4 show results in the "shock-control" format; hence a positive value means that the variable in question is larger with the pipeline and gas developments, plus accommodating policies, than without them.

In some earlier studies of the pipeline, we assessed the impact of the pipeline and gas production without any offsetting policies. The results showed substantial induced increases in other investment, employment, and so on, all of which were more than fully reversed when the boom was over. In a model like RDX2, with its complicated dynamic structure, a project as large as a Mackenzie Valley pipeline can induce an extended series of cyclical responses, reaching magnitudes that would be intolerable to poli-cy-makers.[8] It may therefore be more realistic to assume that if the federal authorities were willing to approve the project at all, they would also be willing to adopt accommodating monetary and fiscal policies that would cushion but not eliminate the macroeconomic effects of the pipeline. The extent to which the policy-makers would attempt to offset the indirect effects of pipeline construction and operation would naturally depend on the inflation and unemployment conditions anticipated about 1980. That is too far ahead for business cycle prediction, so we have chosen offsetting policies without special regard for business cycle conditions, and have established a control solution with fairly even evolution of employment and prices through the mid-1980's.

The policy offsets underlying Tables 3 and 4 are:

(1) The corporation income tax rate is raised from 43.5 to 44.5 per cent in 1975 and 1976, lowered to 42.5 per cent in 1979 and 1980, and is elsewhere equal to its control solution value of 43.5 per cent.

(2) Personal income tax rates (federal and provincial) are 1.05 times their control values in 1976 and 1977; .95 times their control values between the first quarter of 1978 and the fourth quarter of 1981.

(3) Monetary policy is accommodating by raising the target short-term interest rate during construction and lowering it afterwards, and increasing bank liquidity during con-struction (details in Appendix C).

Table 3 shows the results for income, expenditure, prices, employment, and migration. The policy offsets we have used cushion but do not eradicate the basic pattern of relative boom during construction and relative slack afterwards. This is due to the basic multiplier and accelerator mechanisms that influence real expenditures and financial flows. In response to our earlier simulations, which showed very large induced expenditure effects, it was suggested that the temporary nature of the investment project might dampen the accelerator influences. To model this possibility, our 'preferred output' measures are increased in 1975, and less thereafter, to dampen the extent to which firms would choose to expand their plant capacity to meet demands that are thought temporary.

TABLE 3
Macroeconomic Effects of Pipeline & Gas Production
with Policy Offsets
(Annual Values: Shock Results minus Control Results)

Year	Gross National Expenditure (Current $ Millions)	Business Output (1961 $ Millions)	Consumer Price Index (1961 = 1.0)	Number Employed Millions	Unemployment Rate (%)
	YGNE	UGPP	PCPI	NE	RNU
1973	40	15	-0.000	0.001	-0.006
1974	74	33	0.000	0.002	-0.015
1975	1785	635	0.000	0.029	-0.241
1976	3029	1076	0.003	0.075	-0.631
1977	3053	685	0.011	0.074	-0.479
1978	2436	187	0.017	0.041	0.211
1979	185	-868	0.017	-0.016	0.917
1980	-1113	-1264	0.012	-0.060	1.008
1981	-1561	-1066	0.007	-0.071	0.653
1982	-824	-494	0.001	-0.049	-0.021
1983	523	-162	0.000	-0.016	-0.583
1984	2135	-96	0.007	0.004	-0.518
1985	3022	-375	0.020	0.002	-0.105

Year	Investment in Machinery and Equipment	Investment in Non-Residential Construction	Inventory Investment	Aggregate Demand for Business Output	Desired Supply of Business Output
			(millions of 1961 $)		
	IME	INRC	IIB	UGPPA	UGPPD
1973	3	1	-1	19	2
1974	7	5	2	37	12
1975	115	47	-97	829	114
1976	126	132	14	1210	356
1977	73	182	92	593	548
1978	-120	9	3	32	723
1979	-336	-308	38	-1181	499
1980	-447	-553	-131	-1384	-136
1981	-365	-626	-133	-977	-713
1982	-37	-412	125	-353	-1028
1983	238	-58	183	-68	-1032
1984	275	150	-34	-25	-621
1985	212	123	-130	-409	-211

Year	Total Wages	Corporate Profits	Personal Income	Disposable Personal Income	Population (Millions)
			(Millions of Current Dollars)		
	YW	YC	YP	YDP	NPOPT
1973	5	31	9	9	0.000
1974	21	47	31	29	0.000
1975	417	1368	467	432	0.001
1976	1327	1819	1347	525	0.007
1977	1938	1297	1847	604	0.034
1978	1701	680	1804	2026	0.073
1979	598	-724	940	1712	0.092
1980	-436	-1140	17	1031	0.064
1981	-1042	-1059	-534	784	0.000
1982	-915	-493	-297	422	-0.064
1983	40	-187	756	821	-0.099
1984	1486	-178	2266	1911	-0.085
1985	2457	-313	3307	2600	-0.042

TABLE 4
**Macroeconomic Effects of Pipeline & Gas Production
with Policy Offsets**
(Annual Values: Shock Results minus Control Results)

Year	Trade Account Impact of Pipe and Gas	Trade Balance	Capital Account Impact of Pipe and Gas (millions of current dollars)	Balance Of Payments	Price of Foreign Exchange $C/$US
	XBALP&G$	XBAL$	FBALP&G	UBAL	PFX
1973	-14	-21	26	-4	0.000
1974	-14	-31	30	6	-0.000
1975	-623	-916	597	-389	0.005
1976	-641	-1067	927	-89	0.008
1977	-507	-907	905	339	-0.001
1978	-251	-535	857	587	-0.012
1979	136	171	329	444	-0.023
1980	457	831	-101	65	-0.014
1981	585	1102	-222	-19	-0.003
1982	596	564	-235	-122	-0.010
1983	528	-479	-136	-434	-0.026
1984	527	-1075	-136	-750	-0.030
1985	566	-926	-136	-495	-0.014

Year	Imports of Goods And Services From The US	Exports of Goods and Services To The US (millions of current dollars)	Exports of Goods and Services To Other Countries	Total Imports	Total Exports (millions of 1961 dollars)
	M$12	X$12	X$13	M	X
1973	5	0	0	5	-0
1974	11	0	-2	13	-0
1975	238	27	12	190	1
1976	407	98	5	290	23
1977	313	26	-33	288	-23
1978	86	-77	-66	260	-64
1979	-334	-241	-113	48	-131
1980	-550	-229	-21	-231	-109
1981	-466	-65	56	-305	-15
1982	-114	-90	-105	87	-40
1983	254	-309	-416	639	-193
1984	452	-427	-558	851	-285
1985	421	-387	-461	698	-257

Year	Liquid Asset Ratio of Banks (%)	Interest Rate on 0-3 Year Government Bonds (%)	Chartered Bank Deposits at the Bank of Canada (Mill. of Current $)	Supply Price of Capital to Business (%)	Federal Govt. Income-Expend- iture Balance (Mill of Current $)
	RABEL	RS	ABBCD	RHO	GBALF
1973	0.062	-0.009	-0	0.004	8
1974	-0.054	0.020	-0	0.016	13
1975	-0.040	-0.013	30	0.193	382
1976	-0.502	0.453	-8	0.538	1250
1977	-0.436	0.215	25	0.649	1526
1978	-0.060	0.156	78	0.752	-121
1979	0.520	-0.347	136	0.538	-1014
1980	1.244	-0.824	219	0.258	-1234
1981	0.824	-0.632	184	-0.088	-1430
1982	-0.258	-0.150	56	-0.319	-721
1983	-0.754	-0.016	16	-0.375	-21
1984	-0.694	0.090	20	-0.250	168
1985	-0.625	0.161	7	-0.040	189

Note the extent to which the personal tax changes make the swings in personal disposable income (YDP) smaller than those in personal income (YP). Total population (NPOPT) is different by the amount of accumulated net migration induced by the investment surge and consequential slack.

Table 4 shows the consequences for trade, capital flows, financial conditions, and the federal government balance. The direct trade (XBALP&G$) and capital flow (FBALP&G) impacts of the project are repeated from Table 2 for easier comparison with the overall changes in the trade surplus (XBAL$) and the net overall balance on trade and capital accounts (UBAL). It should be noted that the results for the total balance of payments (UBAL) and the balance on trade account (XBAL$) include both direct and induced effects of the project, while the trade categories (M,M$12, etc) contain only the induced effects.

During construction, the induced trade changes amplify the direct trade deficit for the project. During the operating phase, the lower level of activity and the higher value of the Canadian dollar combine to reduce other exports to make room for the pipeline tariff and gas export receipts. The overall long-term balance of payments is negative during early construction and after 1980. The cyclical movements of the price of foreign exchange are shown by the results for PFX in Table 4.

The dynamic simulations are not the best source of information about the equilibrium exchange rate change required to accommodate the pipeline and gas exports. In the published description of RDX2, there is a table of exchange rate effects[9] on trade flows. Those values suggest that to reduce trade flows (ignoring all but relative price effects) to offset the 1981 net direct pipeline and gas export balance of payments effects (XBALP &G$ + FBALP&G) of $255 million, the price of foreign exchange (PFX) would have to be reduced by about 1½ per cent, taking account of the fact that 1981 prices are assumed to be about 65 per cent higher than in 1968.

Alternatively, of course, the price of foreign exchange would not have to be changed at all if Canadian dollar prices of exports and import substitutes were adjusted upwards by 1½ per cent. This kind of calculation is more revealing than the exchange rate results in Table 4 because the exchange rate results from the simulations are very dependent on the cyclical response to the project, and the cyclical response is itself heavily influenced by the particular policy offsets assumed.

D. SUMMARY AND QUALIFICATIONS

In this chapter, we have presented detailed quantitative models of a Mackenzie Valley pipeline and of related gas production in the Mackenzie Delta. In Chapter 10, these models are supplemented by equations calculating the economic costs and benefits of the proposed project compared to alternative uses for our natural and financial resources. In the present chapter, the model of pipeline financing and operation has been used mainly to illustrate some problems in the tariff-setting procedures used by the National Energy Board, and to assess the general or macroeconomic repercussions of the pipeline and gas production.

The problems with the NEB tariff-setting procedure are mainly due to the methods used for allocating over time the return on the capital invested in the pipeline. By inadequately treating inflation, and by allocating capital charges according to the depreciated value of the pipeline, the NEB procedure overcharges the early users relative to the later users of the pipeline.

The main contribution of the macroeconomic simulations is to illustrate the pattern of induced business cycles set in train by a large investment project. These results enable us to assess the macroeconomic statements made by Earle Gray in Chapter 2 (leaving until Chapter 10 the consideration of the other advantages claimed for the Mackenzie Valley pipeline proposal). In Chapter 2, emphasis is placed on three possible macroeconomic consequences: a "major contribution to national income" during construction; the long-run impetus to employment in Alberta and the north; and the generation of a trade surplus during the operations phase. I shall deal with these issues in turn.

It is clear from our simulations that the induced boom in expenditure and employment is large in relation to the direct expenditure and employment on pipeline construction. Our results also show a large induced slump after the construction is over. There is considerable reason for doubting whether the construction phase of a Mackenzie Valley pipeline could ever be expected to improve, on balance, Canada's macroeconomic performance. The labour and material requirements for arctic pipeline construction are so specialized, and the lead time so long, that a Mackenzie Valley pipeline would be inefficient and unwieldy as a make-work project. In addition, the large-scale and temporary nature of the construction phase impose adjustment costs not directly borne by the builders of the pipeline, and not adequately captured by macroeconomic simulations.

Gray's second macroeconomic point is that the development of the Mackenzie Delta gas resources would provide impetus for long-term increases in employment and activity in the north. From the national point of view, such a shift of our human resources has a net advantage only if the project in question is more attractive than available alternatives in other regions. In the case of non-renewable resources, as emphasized in Chapter 10, we also have to make sure that we do not undervalue the potential returns from deferred development. Our analysis in that chapter also shows that premature use of high cost arctic gas would reduce the incomes and economic rents from Alberta gas. Thus, incomes would be generated in the north at least partially at the expense of future incomes everywhere and present incomes in southern Canada.

The third macroeconomic point from Chapter 2 is that the several hundred million dollars annual export surplus (XBALP&G$ in Table 4) would help to offset what might otherwise be a substantial trade deficit. Under a flexible exchange rate, such a point is irrelevant, as the exchange rate always adjusts to keep the sum of trade and capital flows in balance. If there would have been a trade deficit in the 1980's without Delta gas, under a flexible exchange rate, then there would also be capital account surplus. To fit in the extra exports, capital inflows or other exports would have to be cut back, and imports increased, by amounts equal in total to XBALP&G$. Whether we would be better off to substitute the pipeline tariffs and gas exports for other items in the balance of payments must depend on the net long-term benefits from the various trade alternatives. The only exception to this general point arises if a temporary export possibility promises to offset a temporary decline in other exports, thus aiding the short-term stability of the balance of payments and the foreign exchange market. It is clear that a Mackenzie Valley pipeline cannot be thought of in such terms.

Having dealt with the macroeconomic points raised in Chapter 2, it is appropriate to consider how the empirical measures of the macroeconomic impact of a Mackenzie Valley pipeline might alter if the development were deferred in the manner suggested by cases 2 and 3 in Chapter 10. The deferred projects described in Chapter 10 differ in several respects from the proposal (Case 1) analysed in this chapter. Cases 2 and 3 involve five and ten year deferrals respectively. Deferral would reduce the relative macroeconomic impact of pipeline construction, because of the expected continuation of growth in the supplies of capital and labour. Neither case 2 nor case 3 involves any export of Canadian gas from the Mackenzie Delta, and only case 2 involves trans-shipment of Prudhoe Bay gas. Thus the expected export surplus during the operating phase would be smaller, or even negative because of interest payments abroad. As time passes, Canada will become less dependent on foreign capital, so that a higher proportion of a deferred project would likely be financed from Canadian sources. It is plausible that almost all the equity and as much as one-half of the debt capital would come from Canadian sources. As in the case of the Gas Arctic proposal, efforts would no doubt be made in the deferred cases to match imports of capital to imports of goods so as to avoid large net balance of payments impacts during the construction phase.

My conclusion, finally, is that attempts to justify a Mackenzie Valley pipeline proposal in terms of macroeconomic employment or balance of payments benefits are essentially on the wrong track. It is necessary to show that such a project could be digested, and that macroeconomic adjustment costs are taken into account when assessing net benefits. Beyond that, any attempts to justify the project on grounds of macroeconomic balance are vitiated by the presumption that there are available a number of more efficient ways of generating quick jobs in the short-run, or stable and satisfying jobs in the long-run.

FOOTNOTES

1. The earlier stages of the pipeline research, and the March presentation to the U.B.C. seminar, were the joint effort of a group of fourth year economics majors: Mary Ann Cummings, Lindsay Gordon, Bruce Havercroft, Jaroslav Naprstek, Bill Pettit, and John Thompson. Don Weisbeck worked on the gas production sector for his graduating essay. John Lester and Robert McRae provided extensive computing assistance throughout the project. Hartley Lewis and Tom Maxwell made many helpful comments. Thanks are also given to our various sources of fact and opinion about natural gas development in Canada's North.
2. The model and some of its properties are described in John F. Helliwell, Harold T. Shapiro, Gordon R. Sparks, Ian A. Stewart, Frederick W. Gorbet, and Donald R. Stephenson, *The Structure of RDX2* (Ottawa: Bank of Canada, 1971). (Bank of Canada Staff Research Studies, No. 7).
3. This construction schedule corresponds with Gas Arctic plans as they were reported in the spring of 1973. These plans were altered

slightly in the summer of 1973 to permit construction to start in mid-1976 with some gas still expected to flow before the end of 1978. We have not worked out the detailed expenditure consequences of this change in plans because we expect still further changes to be made before the proposal is presented to the NEB in late 1973 or early 1974. By that time, the proposed time for completion may also have been postponed, in which case the project may very well look like the one described in this chapter, with all dates being one year later than assumed by our calculations.

4. The values reported for TARIFF and PGASDEL are those for the last quarter of each year, to avoid the appearance of a spurious 1978 average created by the mid-year starting date.

5. A recent statement of the National Energy Board's ambivalence, and of the pipelines' viewpoint, about the allowability of income taxes may be found in *In the Matter of Application Under Part IV of the National Energy Board Act (Rates Application-Phase I) of Trans-Canada Pipelines Limited* (Ottawa: Queen's Printer, 1971).

6. This result also shows up in the pro-forma tariff estimates included in *Arctic Oil Pipe Line Feasibility Study 1972,* prepared by Mackenzie Valley Pipe Line Research Limited, p. 108.

7. Initial versions of the equations and assumptions used in this section, and reported in Appendix B, were presented by Don Weisbeck in his 1973 U.B.C. graduating essay, "Taxation Aspects of Production and Transportation of Northern Natural Gas."

8. The dynamic adjustment mechanisms in RDX2 are more detailed and closely specified, especially for investment, employment, financial markets, and capital flows, than those of the Canadian annual models (TRACE and CANDIDE) also being used to study the macroeconomic effects of northern energy developments. It will be interesting to compare results from the different models: but this cannot be done until the assumptions and methods used in the other studies are available for inspection. In the meantime, I would guess that RDX2 exaggerates the cyclical dynamics of the economy, while the other two models are likely to understate them. These differences would produce different results for investment projects without policy offsets, but need not necessarily alter the size of the indicated policy offsets.

APPENDICES TO
CHAPTER 8

Appendix A.: Mackenzie Valley Pipeline – Equations and Assumptions

Appendix B.: Mackenzie Delta Gas Production – Equations and
Assumptions
Appendix C.: Macroeconomic Simulations – Methods and
Assumptions

APPENDIX A
MACKENZIE VALLEY PIPELINE – EQUATIONS AND ASSUMPTIONS

The appendix has four sections:
1. equations
2. endogenous and exogenous variables
3. data sources and assumptions for exogenous variables
4. Coefficients – values and definitions. In the equations, the coefficients are represented by four-digit numbers preceded by an 'A'

1. EQUATIONS: (IN ALPHABETICAL ORDER OF THE ENDOGENOUS VARIABLES)

Interim loans to the pipeline from Canadian chartered banks
$$ABLPL = FINREQ - JW(FINREQ) + JIL(ABLPL)$$
Set to provide a positive but constant cash balance after all loans have been repaid (JIL is a one-quarter lag)

Accumulated pipeline losses for income tax purposes
$$APL = -YPT \text{ (only if YPT is negative)}$$

Current pipeline non-wage expenditure
$$CPNWE\$ = A1871(PEXOG)$$

Pipeline financing – foreign bond issues
$$FINIPB12 = (A1865)(A1866(JW(FINREQ)))$$
(JW = a 4-quarter moving average starting in the current quarter. New bond and share issues take place only if JW(FINREQ) is positive)

Financing requirements

$$\text{FINREQ} = \text{IPC\$} - \text{YPT\$} + \text{NPO(WQPO)} + \text{CPNWE\$} +$$

$$(\text{MINT \$}/1. - \text{A1869}) + \text{YINTP})(\text{QSTART}) + \text{TCP}$$

$$+ \text{YDIVP} + \text{MDIVP \$}/1. - \text{A1882})$$

Pipeline financing – foreign portfolio share issues
  FIPVP = (1-A1865)(A1867)JW(FINREQ) if JW(FINREQ) is positive
       = (A1867)(FINREQ) (if neither FINREQ nor LPB12 is positive)
Retirement of foreign-held pipeline debt
  FIRETP12 = (A1866)(FINREQ) (only if JW(FINREQ) is not positive)
Pipeline construction investment

$$\text{IPC} = \frac{\text{IPC\$}}{\text{PIPC}}$$

Rate base value of pipeline
(a) Up to and including 4Q78 KPLBASE = JIL

$$(\text{KPLBASE}) + \text{IPC\$} + \text{QSTART}$$

$$[\text{YINTP} + \text{MINTP \$}/(1. - \text{A1869})]$$

(b) From 1Q79 onward KPLBASE = JIL(KPLBASE) − A1880
    (KPLBASE4Q78)
Value, for income tax purposes, of capitalized pipeline investment expenditure
  KPL\$H = (1-A1862)JIL(KPL\$H) + IPC\$
(Depreciation starts when gas flow commences)
Pipeline financing – foreign bond liability
  LPB12 = JIL(LPB12) + FINIPB12 + FIRETP12

Pipeline financing – domestic bond liability
  LPDB = A1865(1-A1866)JW(FINREQ) + JIL(LPDB) − URET

Pipeline financing – book value of domestic equity
  LPDS = (1-A1865)(1.-A1867)[JW(FINREQ)] +
        (1.−A1867)(LPES-JIL(LPES)) + JIL(LPDS)
  unless all pipeline debt has been retired, in which case
  LPDS = (1-A1867)(FINREQ)+JIL(LPDS)

Pipeline earned surplus

$$\text{LPES} = \text{JIL(LPES)} + \text{YPBOOK} - \text{YDIVP} - \frac{1}{1.-\text{A}1882}(\text{MDIVP\$})$$

Pipeline financing – foreign equity at book value

$$\text{LPS}12 = \text{JIL(LPS}12) + \text{FIPVP} + \text{A}1867(\text{LPES} - \text{JIL(LPES)})$$

Pipeline foreign dividends

$$\text{MDIVP\$} = (\text{A}1863)(\text{A}1867)(\text{YPBOOK})(.1.-\text{A}1882)$$
(only if LPES and YPBOOK are positive)

Foreign interest payments

$$\text{MINTP\$} = \text{A}1868(\text{LPB}12)(1-\text{A}1869)$$

Pipeline tariff, in cents per thousand cu. ft. (¢/mcf), excluding cost of transmission gas. Defined only after gas flow starts (when QSTART = 1.) Set equal to .01 (KPLBASE) from 2Q78 TO 4Q78.

$$\text{TARIFF} = [\text{YINTP} + \text{MINTP\$}/(1.-\text{A}1869) + \text{TCP} + (\text{NPO})(\text{WQPO})$$
$$+ \text{A}1880\,(\text{KPLBASE4Q78}) + \text{CPNWE\$} +$$
$$\text{A}1883(\text{KPLBASE-LPDB-LPB}12\text{-ABLPL})]/.9125(\text{A}1881)$$

Accrued corporation tax on pipeline profits

$$\text{TCP} = \text{A}1884(\text{YPT}) \text{ (only if YPT, from equation 2, is positive)}$$

Retirement or sinking fund purchase of domestically held pipeline debt

$$\text{URET} = (\text{A}1866-1.0)(\text{FINREQ})$$
(only if interim bank financing is all paid off and LPDB still positive)

Total receipts from pipeline tariffs

$$\text{YPT\$} = (\text{TARIFF})(\text{EGASFLOW})(.9125)$$

Pipeline domestic dividends

$$\text{YDIVP} = (\text{A}1863)(1\text{-A}1867)(\text{YPBOOK})$$
(only if LPES and YPBOOK are positive)

Domestic interest payments

$$\text{YINTP} = \text{A}1870(\text{LPDB}) + \text{A}1879(\text{ABLPL})$$

Book value of pipeline profits, after taxes, from 1Q79 onward

$$\text{YPBOOK} = \text{YPT\$} - \text{A}1880(\text{KPLBASE4Q78}) - \text{NPO}(\text{WQPO}) -$$

$$\text{CPNWE\$-MINTP\$}/(1.-\text{A}1869) - \text{YINTP} - \text{TCP}$$

2. ENDOGENOUS AND EXOGENOUS VARIABLES

All variables whose names start with A, F, L, or Y, or end with $ are measured in millions of current dollars. The units for other variables are included with the definitions. The endogenous variables are listed first, followed by the exogenous.

Endogenous Variables (in alphabetical order)

ABLPL   Loans outstanding from Canadian banks to the pipeline – used to provide interim financing

APL   Accumulated pipeline losses calculated according to income tax regulations

CPNWE$ Current pipeline non-wage operating expenditure

FINIPB12   New pipeline borrowing abroad – including bonds, term loans from suppliers, foreign banks, etc.

FINREQ   Financing requirements: cash flow before issue or retirement of securities

FIPVP   New foreign issues of pipeline shares (retirements if negative)

FIRETP12   Retirement or sinking fund purchase of foreign-held pipeline debt (negative values)

IPC   Pipeline construction investment, in millions of 1961 dollars

KPLBASE   Rate base value of pipeline

KPL$H   Value, for income tax purposes, of capitalized pipeline investment expenditures

LPB12   Foreign bond liabilities of pipeline

LPDB   Domestic bond liabilities of pipeline

LPDS   Domestic equity in pipeline, at book value

LPES   Earned surplus of pipeline

LPS12   Foreign equity in pipeline, at book value

MDIVP$ Pipeline foreign dividends, after withholding tax

MINTP$ Pipeline foreign interest payments, after withholding tax

TARIFF   Pipeline tariff rate, excluding gas used in transmission, measured in cents per thousand cubic feet (¢/mcf) of delivered gas

TCP   Corporation taxes paid by pipeline

URET   Retirement or sinking fund purchase of domestically-held pipeline debt

YDIVP   Pipeline domestic dividends

YINTP   Pipeline domestic interest payments

YPBOOK   Pipeline book profits

YPT$ Pipeline tariff revenues

Exogenous variables (in alphabetical order)

EGASFLOW   Pipeline throughput, in billions of cubic feet per day of delivered gas
IPC$ Pipeline construction investment
NPC   Employees in pipeline construction, in millions of persons
NPO   Employees in pipeline operation, in millions of persons
PIPC   Price index for pipeline construction
QSTART   Variable with value 1.0 after gas starts to flow
WQPC   Quarterly wage rate for pipeline construction, in dollars/quarter
WQPO   Quarterly wage rate for pipeline operation

3. EXPLANATION OF SOURCES AND ASSUMPTIONS FOR EXOGENOUS VARIABLES

EGASFLOW Pipeline throughout
This series is measured in billion cubic feet (b.c.f.) of gas delivered per day at the southern ends of the pipeline. It starts at .4 in 78Q2 and 78Q3, and grows thereafter by .4 each quarter, reaching full capacity throughput of 4.0 bcf/d in 80Q4.

IPC$ Pipeline construction investment, in millions of current dollars. This variable represents the investment in pipeline construction in current dollar terms. There are six factors making up IPC$

(a) The cost of pipe – The length of the pipeline is assumed to be 2,400 miles. Pipe costs average about $350 per ton. The weight per mile can be calculated[2] to be 1,042 tons/mile. This comes to an approximation of $875 million for all the pipe. Building the pipeline over a three-year period necessitates having a stockpile of pipe before construction begins. It is assumed that pipe will be bought first in 1975 Q1 and then at a constant rate until 1977 Q4. This means that $72.9 million per quarter will be spent on pipe.

(b) Transportation costs – Estimates of transport costs have been seen to be about 10% of total cost.[3] As the total cost is over $5 billion, an estimate of $600 million was made for transport costs. In the breakdown of the $600 million by quarter the following assumptions were made:
1) There will be stockpiling of materials before actual construction starts; 2) There will be a rise in the amount of men

and materials shipped on the MacKenzie starting one quarter before the project starts and ending at the end of 1976 Q4;
3) The amount of transport done will remain constant after 1976 Q4, at a lower amount.

Using the previous assumptions transportation costs were assumed to be $30 million for Q1 and Q2 of 1975, $50 million/quarter for 75Q4, and $30 million/quarter for 77Q1 to 78Q4.

(c) Other construction costs – These costs include the costs of construction equipment and supplies, engineering costs, compressor station costs, research costs and right-of-way costs. Although no precise figures are available, estimates of these costs are as follows:

construction equipment and supplies	$900 million
engineering	$200 million
compressor stations[4]	$800 million
right-of-way and research	$202 million
	$2,102 million

These costs are faced at various times. The research costs and right-of-way costs will have to be covered before construction can begin. In the three quarters before construction starts there will have to be buying of equipment in preparation for construction. In the early building stages more construction equipment and materials will have to be purchased so that near the end of project construction costs will be declining. The following breakdown of costs shows these trends:

1973	Q1	$6.5 million	1975	Q1	$150.0 million
	Q2	$6.5 million		Q2	$150.0 million
	Q3	$6.5 million		Q3	$150.0 million
	Q4	$6.5 million		Q4	$150.0 million
1974	Q1	$6.5 million	1976	Q1	$150.0 million
	Q2	$6.5 million		Q2	$150.0 million
	Q3	$6.5 million		Q3	$150.0 million
	04	$6.5 million		Q4	$150.0 million
1977	Q1	$100.0 million	1978	Q1	$150.0 million
	Q2	$100.0 million		Q2	$150.0 million
	Q3	$100.0 million		Q3	$150.0 million
	Q4	$100.0 million		Q4	$150.0 million

Construction of many of the compressor stations would take place after the gas started to flow in 1978.[5]

(d) Expenditure for direct employment in pipeline construction is equal to (NPC)(WQPC)

(e) Financing costs are assumed to be $105 million, equal to 2% of the value of bonds and share issues. This percentage is based on flotation costs for bonds issued by Trans Canada Pipeline between 1964 and 1971.

(f) Interest paid prior to 2Q78 is included in IPC$. Starting in 2Q78, interest is implicitly included in YPT$. In terms of national accounting theory, according to the UN standard classifications, financing costs, and probably interest paid during construction, should be intermediate consumption rather than investment expenditure. However, the Canadian national accounts do include interest paid during construction as part of investment expenditure, although the magnitude of the item is not known. Most firms have tax reasons for writing off interest payments during construction (at least for own-account construction). The reasoning may be different for regulated utilities, whose capitalized expenditures form the base for allowed rates of return. We think that our present procedure overstates IPC$ somewhat, but the consequences of this are reduced by our separate accounting for pipeline assets and supply capacity.

NPC   Number employed in pipeline construction
This variable represents the number of men to work on the actual building of the line. The total number of employees has been quoted several times in the 7,000[6] range. We have used a value of 7,250 employees during the twelve quarters of construction. There will also be some employment of workers for the three pre-construction quarters and the three post-construction quarters. This value was estimated to be 2,000 during each of these six quarters. Our estimate of construction employment has a slightly lower peak than that of 8,000 reported by Earle Gray in Chapter 2. It would appear that Gas Arctic now plans a later start and slightly more concentration on winter construction than is implied by our figures. Gray's 'average' estimate of 5,000-6,000 is substantially below the 8,000 peak. Perhaps he is averaging over a 5-year construction period.

NPO   Number employed in pipeline operation
This variable represents the number of people required to operate the pipeline. We use an estimate of 600 employees[7] from 1978 Q2 until the end of the life of the pipeline.

PIPC   Price index for pipeline construction
This is a price index variable which has a value of 1.00 in 1961 and is inflated at 5% per year over the construction of the pipeline. This index is used to determine the investment in pipeline construction in 1961-dollar terms.

QSTART
This variable takes the value 1.0 wherever the pipeline is in partial or full operation. It is 0 until and including 78Q1, and 1.0 thereafter.

WQPC   quarterly wage in pipeline construction
This variable represents the average quarterly wage paid to construction workers.

The following estimates have been given:
Welders	$5,000 per month
Equipment Operators	$4,000 per month
Labourers, Truck Drivers	$2,000 per month[8]

From these figures an average wage per man per month of $3,666 was estimated, assuming that there would be more labourers and truck drivers. This average wage is equal to $11,000 per quarter.

WQPO   Quarterly wage in pipeline operation
This variable represents the wages paid to people operating the pipeline. The salary of those employed in NPO is assumed to be about $6,000 per quarter in 1978 Q1, and to grow thereafter at 7% per year.

4. COEFFICIENTS – values and definitions

Number	Value	Meaning
A1862	.01534	Quarterly proportional equivalent of the 6% pipeline capital cost allowance rate permitted for tax purposes
A1863	1.0	Pay out ratio for after-tax book profits, applicable only when earned surplus is positive
A1865	.8	Ratio of debt/(debt+equity) for new long term financing
A1866	.65	Ratio of U.S. to total new pipeline bond issues

A1867	.49	Foreign (U.S.) ownership ratio for pipeline equity
A1868	.0225	Quarterly proportional interest rate on U.S. bond issues
A1869	.15	Rate of withholding tax on interest payments abroad
A1870	.0225	Quarterly proportional interest rate on Canadian bond issues
A1871	6.028	1961$ value of pipeline nonwage operating expenditure
A1872	.1	Import content (from third countries) of pipeline nonwage operating expenditure
A1873	.2	Import content (from U.S.) of pipeline nonwage operating expenditure
A1874	.15	Import content (from third countries) of pipeline nonwage construction expenditure
A1875	.4	Import content (from U.S.) of pipeline nonwage construction expenditure
A1876	3851.	1961$ value of quarterly pipeline construction wage
A1877	1.85	Background simulation value of PGNE in 1978
A1878	9.46	Coefficient applied to (KPL\$H/KB\$) as an additional term in the supply price of capital equation. It is derived from the coefficient in the relative supply variable in the VKB equation of RDX2.
A1879	.02	Quarterly proportional interest rate on interim bank loans
A1880	.0075	Quarterly rate of straight-line depreciation (equal to 3% per annum) used in calculating pipeline book profits and in depreciating the rate base.
A1881	4.0	Pipeline throughput, when fully loaded, in billions of cu. ft. delivered per day
A1882	.1	Rate of withholding tax on pipeline dividend payments abroad
A1883	.035	Quarterly proportional after-tax rate of return allowed on equity-financed portion of pipeline rate base.
A1884	.46	Rate of corporation income tax on taxable pipeline profits

FOOTNOTES TO APPENDIX A

1. A lower estimate of $290 per ton has been made by M. Dimentberg in, "Comparison of Liquid and Vapour Phase Transmission", Liquefaction Ltd., Winnipeg, p. 11. Our initial estimate was based on an average cost of $325 per ton south of 60° and $375 per ton north of 60°. These figures are very uncertain, as 1976 quotes for pipe of the required size and quality are understandably difficult to obtain. Industry estimates in mid-1973 suggest that quotations may run over $450 per ton. On the other hand, Gas Arctic's most recently proposed system (as described in the revised chapter by Earle Gray) involves slightly fewer tons of pipe, because the two southern branches are of 42" pipe. Our pipe cost estimate of $875 million amounts to about $370 per ton of pipe laid over 2403 Canadian miles of the route described by Gray.

2. Calculations:
   Weight of pipe/mile = wt/ft × 5280 ft/mile
   wt/ft = .778/.50 × 253.65* = 394.68 where 253.65 lbs/ft is the weight of 48" pipe, with wall thickness of .5 inches and where .778/.5 finds the weight of 48" pipe, with wall thickness of .778".
   *from: CSA Standard Z245.1, Table 4, p. 39, General Requirements for plain-end welded and seamless steel line pipe.
   Weight per mile = 1,041.96 tons/mile
   2400 miles × 1,041,96 tons/mile × $350 = $875.25 million
   We have not altered these calculations to reflect the planned use of 42" pipe over some of the 2400 miles of the route proposed by Gas Arctic in mid-1973. Pipe costs are rising fast enough, because of shortages and general price increases, that we now think our $875 million estimate of pipe costs is more likely to be too low than too high. See the preceding footnote for more details.

3. From: David Crane, "MacKenzie Gas Pipeline – an Engineer's Dream and a Builder's Nightmare," *Toronto Star* (September 26th, 1972), p. 4.

4. The total figure for compressor stations is made up of two components. Those stations in permafrost areas will also have refrigeration units to chill the gas for transport.

	# of stations	Cost $/station	Cost
Permafrost area with refrigeration	20	$ 25 m.	= $500 million
Non-permafrost area without refrigeration	20	$ 15 m.	= $300 million
			$800 million

5. From: Earle Gray's presentation to the UBC seminar, March 13, 1973 and his chapter in this volume

6. From: Earle Gray, Statements at an education seminar in Inuvik, December 5, 1972; and David Crane, *op. cit.*, p. 4

7. A lower estimate of 200 northern jobs is from p. 14 of Earle

Gray's paper presented to the UBC seminar on March 13. Our overall estimate of 600 jobs parallels that in the *MacKenzie Valley Oil Pipeline Feasibility Study 1972* prepared by Mac-Kenzie Valley Research Ltd., and is similar to the more recent Gas Arctic estimate reported in Chapter 2.
8. From: "Arctic Pipelines – Catalyst for Northern Development," *Canadian Business Magazine* (November, 1972), p. 57.

APPENDIX B
MACKENZIE DELTA GAS PRODUCTION – EQUATIONS AND ASSUMPTIONS

The appendix contains:

1. equations
2. endogenous variables
3. exogenous variables
4. coefficients – values and definitions.

1. EQUATIONS

The equations are primarily accounting relationships designed to measure and allocate the profits from gas production. Each four-digit number preceded by A is a coefficient, for which the value and meaning are to be found in the fourth part of this appendix. JIL refers to a one-quarter lag. The two equations showing the balance of payments impacts of the pipeline and gas production combined are explained in Appendix C.

$$\text{CGASP\$} = (\text{EGASD}/\text{A}1897)(\text{PEXOG})(\text{A}1898)$$

$$\text{KEDP\$} = \text{JIL} (\text{KEDP\$}) + \text{IGPC\$} + \text{IGWC\$} + \text{EXPLORE\$}$$

$$\text{KGAS} = \text{JIL} (\text{KGAS}) + (\text{IGWC\$} + \text{IGPC\$})/\text{PEXOG} - \text{A}1896 \\ (\text{EGASD})$$

$$\text{KGP\$H} = (1.0 - \text{A}1862) \text{ JIL } (\text{KGP\$H}) + \text{IGPC\$}$$

$$\text{KGW\$H} = (1.0 - \text{A}1893) \text{ JIL } (\text{KGW\$H}) + \text{IGWC\$}$$

$$\text{KGP\$HB} = \text{JIL} (\text{KGP\$HB}) + \text{IGPC\$} + \text{IGWC\$} - \text{A}1885(\text{EGASD})$$

$$\text{PGASDEL} = \text{TARIFF} + 1.125(\text{EPGAS}) + \text{A}1899(\text{PEXOG}/1.48)$$
(the .125 takes account of the fact that gas used in transmission is one-eighth of the amount delivered)

$$\text{TROYAL} = \text{A}1886 (\text{YGAS\$})$$
(up to and including 1983 1Q)

TROYAL = A1887 (YGAS\$) (from 1983 2Q)

YGAS\$ = EPGAS(EGASD).9125 (.9125 is the constant necessary to convert (¢/mcf) (bcf/d) into millions of dollars per quarter)

YGASB = YGAS\$ − A1885(EGASD) − CGASP\$ − A1892 (KGP\$HB − LFREE) − TROYAL − EXPLORE\$

Taxable profits, YGAST, are complicated primarily because exploration and development expenditure prior to 1976 may be accumulated and then used as a basis for earned depletion in 1977 and after. We assume that the firms in question are large enough to reap the accumulated benefits during 1977 and 1978, allowing the post-transition system to come fully into effect in 1979.

Until 1976 Q4

YGAST = [YGAS\$ − A1893(KGW\$H) − A1862(KGP\$H) − CGASP\$) − A1892(KGP\$HB − LFREE) − TROYAL − EXPLORE\$](1.0-A1891)

1977 Q1 to 1978 Q4

YGAST = YGAS\$ − A1893(KGW\$H) − A1862(KGP\$H) − CGASP\$ − A1892(KGP\$HB − LFREE) − TROYAL − EXPLORE\$ − .125 (A1891)(KEDP\$4Q76) − A1891 (EXPLORE\$ + IGPC\$ + IGWC\$)

From 1979 Q1.

YGAST = YGAS\$ − A1893(KGW\$H) − A1862(KGP\$H) − CGASP\$ − A1892(KGP\$HB − LFREE) − TROYAL − EXPLORE\$ − A1891(EXPLORE\$ + IGPC\$ + IGWC\$)

2. ENDOGENOUS VARIABLES (in alphabetical order)

CGASP\$ Operating expenses

KEDP\$ Accumulated exploration and development expenditure − used to accumulate earned depletion prior to 4Q76

KGAS   Constant (1961) dollar value of net stock of gas production plant and equipment

KGW\$H   Stock of gas well equipment, valued for capital cost allowance purposes

KGP\$H   Stock of gathering lines and processing plant, valued for capital cost allowance purposes

KGP$HB	Stock of all fixed assets capitalized for book accounting
PGASDEL	Delivered price of gas at a southern terminus of the Gas Arctic system, in cents per thousand cubic feet.
TCGAS	Corporation tax on Delta gas production (can be negative, as the gas production would be undertaken by large integrated firms with profits from other sources)
TROYAL	Federal royalties from gas production
YGAS	Sales of Delta gas (includes gas used up in transmission in the main pipeline, but not in gathering and processing)
YGASB	Book profits from gas production, after tax and royalties
YGAST	Taxable profits from gas production

3. EXOGENOUS VARIABLES (in alphabetical order)

EGASD	Mackenzie Delta gas entering pipeline. In the simulations reported in this chapter we assume that Mackenzie Delta and Prudhoe Bay gas will each provide 50% of the input to the pipeline. EGASD is therefore 2.25 bcf/d when the pipeline is fully loaded.
EGASXD	Proportion of EGASD that is exported. The assumption for EGASXD was 1.0 in earlier versions of the study, but is .5 for the main simulations reported in this chapter.
EPGAS	Wellhead gas price, measured in cents per mcf. Our price assumptions are based on existing contracts covering 10 trillion cu. ft. sold by Imperial Oil to the Michigan-Wisconsin Pipeline Co. and the Natural Gas Pipeline Co. of America. We assume (probably incorrectly) that the prices are in Canadian currency. Quoting from page 3 of the 1972 Imperial Oil card prepared by the Financial Post Corporation Service, and treating 1978 as the first full contract year, we have

1978-79	32.0 ¢/mcf
1980-84	34.0 ¢/mcf
1985-89	39.0 ¢/mcf

1990-94   44.0 ¢/mcf
1995-99   49.0 ¢/mcf

We extended this series beyond the 22-year contract period by raising the price to 54. in 2000 and having it grow at 4% per year thereafter. The above prices are minimum estimates, as the agreement contains a clause stipulating that the buyers must match any higher wellhead prices north of 60°.

EXPLORE$ Exploration and development drilling expenditure. We assume total exploration cost of $112 million, distributed $5 million per quarter in 1973 and 1972, and $2 million per quarter from 1975 to 1983. These figures, based on earlier estimates by Bache & Co., McLeod & Weir, and Wood Gundy, are probably low in the light of the late 1973 report that only 7 tcf of Delta gas has been proven.

IGPC$ Investment in gathering lines and gas processing equipment

IGWC$ Investment in gas well equipment

These two items (IGPC$ and IGWC$) are kept separate because they are treated differently for tax purposes, gas well equipment obtaining 30% and processing equipment 6% capital cost allowances on a declining balance basis. The total of both items, to develop Delta gas in the quantities we are assuming, has been estimated (by the sources noted in the EXPLORE$ definition) to range from $350 million to $1.4 billion. We are assuming $840 million divided as follows, in millions of dollars per quarter

	IGPC$	IGWC$
1975	32.	8.
1976	32.	8.
1977	52.	13.
1978	52.	13.

LFREE   According to the contract described in the EPGAS definition, Imperial Oil receives interest free loans of $10 million per year from 1972 to 1975, $100 million when pipeline construction goes ahead, and $100 million when it is completed. Gulf Oil apparently also has similar arrangements in its sale

contract, covering 4 trillion cu. ft., with Pacific
Gas. In earlier versions of this study, we assumed
that 26 tcf of Mackenzie Delta gas would be
exported, and we made LFREE twice the amount
that Imperial Oil obtained on the basis of a
contract for 10 tcf. The latest Gas Arctic plans, as
reported in Chapter 2, envisage exports of only
10-15 tcf, and our main simulations assume
exports of 25% of the total, or 13.3 tcf. We
therefore assume LFREE to be 30% larger than the
loan arranged by Imperial for 10 tcf.
The resulting series is as follows, in millions of
current dollars:

1973	13
1974	26
1975	169
1976	208
1977	247
1978	312

After 1980, the loans are assumed to be paid off at
the rate of $8 million per quarter.

PEXOG   An exogenous price index taking the value of the
GNE deflator in 1Q73 (1.58) and growing thereafter
at an annual rate of 4%.

PGAS   The cost of the next cheapest substitute in cities
bordering the great Lakes for Mackenzie Delta or
Prudhoe Bay gas. Our initial assumption for this
variable is 100/mcf in 1978, rising at 6% per year
until 2000, and thereafter at 4% per year.

NUMBER	Value	Meaning
A1862	.01534	Quarterly proportional equivalent of the 6% capital cost allowance rate permitted for pipelines and gas processing plants.
A1885	2.886	Quarterly depreciation for book purposes, measured in millions of dollars per bcf/d. When multiplied by EGASD, it gives depreciation in millions of dollars per quarter. It is equal to

$$\frac{\Sigma_{IGWC}\$ + \Sigma_{IGPC}\$}{\Sigma_{EGASD}} \quad (.9125)$$

A1886   .05   Proportion of wellhead price collected as

royalty during the first five years of operation

A1887 .10 Proportion of wellhead price collected as royalty after the first five years of operation

A1888 .43 Weighted marginal rate of corporation income tax on taxable producer profits. According to Bill C-259 the rate for all corporations will drop to 46% in 1976 and thereafter. Under J. Turner's 1972 budget, the rate is 40% for manufacturing and processing. We assume half the profits to be taxable at 40% and half at 46%

A1890 .0283 Average supply price of capital to the Canadian business sector. Equal to .25(.04 + RHOR), where RHOR is the 1956-70 average of the RDX2 supply price of capital in real terms.

A1891 .3333 Proportional rate of depletion allowance for tax purposes.

A1892 .00844 Interest paid on producers' debt, measured as a quarterly proportion of the book value of producers' fixed assets. The assumed borrowing rate of .09 is multiplied by Imperial Oil's 1971 ratio of debt/ (debt + equity) of .375, and the result is divided by 4.

A1893 .085 Quarterly equivalent of the 30% declining balance capital cost allowance rate permitted for gas well equipment.

A1895 .01586 Average tax return on capital invested in typical Canadian industry. Calculated as

$$\frac{.35(\text{indirect taxes} - \text{customs duties}) + \text{corporation ta}}{\text{capital invested}}$$

where .35 is the approximate share of capital's return to total value added in Canadian industry, 1956-70 average data are employed, and the RDX2 variable KB$ is used for 'capital invested', thus providing a conceptual parallel with (KGAS) (PEXOG)

A1896 1.551 Quarterly constant dollar depreciation, in millions of 1961 dollars per unit of EGASD. Similar to A1885, but uses constant dollar totals of investment expenditure.

A1897   2.159   Average extraction rate of gas from the
                    Mackenzie Delta, in billion cu. ft. per day
                    $$\frac{(26.6)\ (1,000)}{(365)\ (33.75)}$$
                    where 26.6 is the total extracted, in trillion
                    cu. ft., and 33.75 is the number of years of
                    extraction.

A1898   14.63   Operating expenses in the typical quarter,
                    measured in millions of 1961 dollars per
                    quarter, in the case where MacKenzie
                    Delta gax provides 50% of the pipeline
                    throughput.

APPENDIX C
MACROECONOMIC SIMULATIONS – METHODS AND ASSUMPTIONS

For brevity, this appendix is cryptic; exact coding is available on request.

1) LINKS WITH THE REST OF THE MODEL

The main links between the PIPE&GAS sector and the rest of the RDX2 model are provided by summary measures of the direct trade (XBALP&G$) and capital flow (FBALP&G) impacts of the pipeline and gas producing activites. There are two definitions for XBALP&G$, because interest payments are part of investment expenditure until the pipeline is brought into service in 1978 Q2. The variables used in the definitions are themselves defined in Appendixes A and B.

Up to 1978 1Q

$\quad$ XBALP&G$ = YPT$ (.5) (1. + EGASXD) + YGAS$ (EGASXD) −
$\qquad$ .35 (IGWC$ + IGPC$)
$\qquad$ − A1873 (CPNWE$ + CGASP$)
$\qquad$ − (A1874 + A1875)(IPC$ − (WQPC)
$\qquad$ (NPC) − MINTP$)

From 1978 2Q

$\quad$ XBALP&G$ = YPT$ (.5) (1. + EGASXD) + YGAS$
$\qquad$ (EGASXD) − MDIVP$ − MINTP$
$\qquad$ − A1873 (CPNWE$ + CGASP$)
$\qquad$ − .35 (IGWC$ + IGPC$)

The capital account variable, FBALP&G, has a uniform definition throughout

FBALP&G=FINIPB 12 + FIRETP 12 + FIPVP + LFREE – JIL (LFREE)

To establish the links, XBALP&G$ is coded into XBAL$ and YGNE, and both XBALP&G$ and F)ALP&G are coded into UBAL 12.

(Model users should note that separate coding into YGNE is necessary because the components of XBAL$, rather than XBAL$ itself, are in the model coding for YGNE).

The important point of distinction between our treatment of the pipeline and of gas production is that value added in the pipeline is treated as the output of a separate sector, while sales of gas are treated as part of general business output. Thus, for example, we do not have to subtract gas production when moving from YGNE to YGPP, and we do not require separate treatment of employment in gas production and processing, because it is explained as part of NMMOB.

The financing of gas exploration and production is also treated indirectly, except for the interest-free loans. For all other financing, retirements, interest and dividend payments at home and abroad, the explanation is provided by the aggregate RDX2 equations, for gas production, while they are the subject of separate equations for the pipeline. Gas production is part of YGPP, so that the profits are part of YC, and so forth. Thus our pipeline and gas production simulations to 2011 do not include interest and dividends on gas production in XBALP&G$, although they are of course included for the pipeline.

On the other hand, our treatment of the pipeline as a separate industrial sector requires NPO and NPC to be added to total employment NE, pipeline wages to be added to YW, pipeline domestic interest and dividends to be added to YP, and value-added by the pipeline to be subtracted in the move from YGNE to YGPP. The details of the coding are available on request.

To capture the direct impact of domestic pipeline financing on the supply price of capital to other Canadian business investment, RHOR is increased by A1878 times the ratio of KPL$H to KB$. A1878 is derived from the coefficient on the relative supply variable in equation 18.1 of RDX2, producing a maximum increase in RHOR of about .2 percentage points.

Finally, bank loans to the pipeline, ABLPL, are subtracted from the equation defining ABEL, thus reducing the banks' earning liquid asset ratio RABEL by the appropriate amount.

2) BACKGROUND SIMULATION TO 1984

We have tried to use simple internally consistent forecasting rules for all exogenous variables using the following pattern:

1)	assumed rate of technical progress (historical rate of growth of ELEFF)=	2.69%
2)	population increase	1.56%
3)	real expenditures	4.25%
4)	inflation	4.00%
5)	money expenditures and assets	8.25%

Some miscellaneous adjustments were made to improve the growth properties of the system. We have removed DSWPB from ULS in simulation, on the grounds that swap deposits determined from the portfolio model should not represent a disequilibrium in desired short term liabilities.

Gross national expenditure in nominal terms grows at an average annual rate of 8.6% over the twelve years 1973-85, while the average rate of growth in the implicit price of GNE is 4.8%. The average unemployment rate is 5.5%. Underlying these averages is some cyclical variation, with more rapid expansion in the mid 1970's and mid 1980's than at the end of the 70's. These variations are not extreme, annual rates of growth in YGNE ranging from a low of 8.1% to a high of 13% (in 1985). In terms of current dollars, YGNE passes 150 billion per year in 1976 and 200 billion in 1980. We have not attempted to make our background simulation an accurate forecast; only to see that it is not totally implausible. We use it only to provide a basis for our shock minus control results.

3) POLICY ADJUSTMENTS

To help cushion the economy we have made the policy changes outlined in the text, effected by:

a) setting RTCA to 44.5 in 1976 and 1977, and to 42.5 in 1979 and 1980;

b) multiplying RTPYF1C to 4C by 1.05 in 1976 and 1977, and by .95 between 1978Q1 and 1982Q1.

c) TEMP in RS equation set to .3 in 1973,-.3 in 1979,-.5 in 1980, and -.3 in 1981.

d) DDGFB raised by 100. from 1976 to 1978 inclusive.

The adjustments to preferred output were made by coding in TEMP's for UGPPANP and UGPPAMP equal to 30. and 40. respectively, throughout 1975. This naturally gives a continuing effect that gradually declines as preferred output is depreciated.

9.
FRONTIER OIL AND GAS EXPLOITATION IN CANADIAN INDUSTRIAL STRATEGY

A. Milton Moore

Initial concern about the proposed Mackenzie Valley gas pipeline centred on its likely impact on the fragile natural environment of the arctic, its effects on northern native people, and its impact on the rest of the Canadian economy during the construction period. More recently, attention has shifted to broader economic concerns: the effects on the Canadian economy of massive diversions of labour and capital into the exploitation of frontier oil and gas resources over a long period. In this paper, I shall argue that the initial concerns are not crucial to the continuing debate because the "correct" solutions to these problems are common to the several alternative approaches to the exploitation of frontier oil and gas. I then discuss what I consider to be the crucial policy issues, which relate to the long term economic impact of development.

This is not to say that the initial concerns, which are discussed in detail in preceding chapters, are not important. They are. And it is to be hoped that public interest groups such as the Canadian Arctic Resources Committee and Pollution Probe at the University of Toronto, will maintain their pressure on the government to ensure that the interests of native people and the need for environmental protection are accorded the weight and priority they warrant. But the "correct" solutions to these problems are those that the majority of Canadians would prefer if they were fully informed, and while I make no claim to special expertise in divining public opinion, I consider that general statements of the correct solutions are not controversial.

With respect to the environmental problems discussed in Chapter 7, the developers should be required to implement safeguards to reduce ecological damage to the point at which the cost of additional protective measures rises above the value of probable damage prevented. With respect to the effects on native people discussed in Chapter 6, the desired solution depends upon notions of fairness and equity. The choices presumably range from full compensation for the lands to which native people claim hereditary rights, to a requirement that they be left better off than if no gas or oil were produced. I would think it unlikely that many Canadians would be comfortable with a decision to impose a net detriment on the northern Indians and Eskimos. On the other hand, I doubt that a majority would recognize a right of the native people to block the production of the petroleum reserves altogether in order to preserve fishing and hunting grounds. But whatever the correct solution to these two issues may be, they will be the same under any of the alternative strategies for development of the frontier resources.

Different interpretations are made of the economic dislocation that would be caused by large pipeline construction. But once agreement is reached on the appropriate methods of measurement and forecasting, there exist standard monetary and fiscal procedures for overcoming obstacles and reducing disturbances to tolerable magnitudes, as explained in Chapter 8. The real question is whether the benefit to the economy over the long run is commensurate with the adjustment costs, and whether the capital devoted to frontier gas and oil for export could be put to a more productive use.

On the basis of the above argument, the crucial issues all relate to the *long term* impact of frontier resource exploitation on the Canadian economy. The remainder of this paper, therefore, is concerned with the relation of frontier resource development to Canadian economic and industrial strategy. Crucial to this is the matter of the timing of frontier oil and gas exploitation – the subject of the following section. Policy affecting the pricing of Canadian oil and gas and the provisions for governmental appropriation of resource values are obviously important in this context also, but these questions are dealt with in detail in other chapters.

A. THE TIMING OF FRONTIER PRODUCTION

Among the more enlightening parts of the 1973 report on energy released by the Minister of Energy, Mines and Resources are the recent estimates of the Geological Survey of Canada (G.S.C.) of potential reserves of oil and gas in the frontier regions, and the estimates of costs of discovery, production and delivery made by the staff of the Department.[1] Table 1 summarizes this information together with similar information about the Athabasca heavy oil sands (also provided in the report). These estimates of potential reserves can be compared with the projected Canadian demand for oil and gas to the year 2010 summarized in Table 2. It should be remembered that the data relating to potential reserves are estimates only. Seismic surveys have been made over much of the Canadian north and much is known about the geological structures of the basins, so the figures are not merely numbers on a page. Nevertheless, great uncertainty remains about the potential of any of the basins – and all of them together – until the search is completed. So far, oil discoveries have not been significant anywhere in the frontier areas and gas in quantities sufficient to justify the transfer of the estimated reserves from "potential" to "probable"[2] has been discovered only in the Mackenzie Delta, the Arctic Islands of the Sverdrup Basin and, perhaps, in the vicinity of Sable Island off the coast of Nova Scotia.

Moreover, a significant probability of oil and gas reserves recoverable at competitive costs is one thing. Establishing that there are reasonably assured supplies sufficient to meet Canadian requirements and existing export commitments each year for a chosen number of years ahead, is another. It is established policy that the National Energy Board (NEB) should not authorize new gas export contracts or the export of any oil from the Western Provinces in any month, unless it is satisfied that the authorized exports are surplus to Canadian requirements. It is possible to challenge the "domestic requirements first" rule as being unprofitably rigid. However, a full discussion of the issue would require a considerable digression from the main concerns of this article. Suffice it to say, therefore, that the rule is accepted by all interested parties, and my own conclusion is that the rule is eminently sensible in the circumstances. However, assigning priority to the security of domestic supplies creates a dilemma: the extent of the frontier reserves remains uncertain until they are discovered, but once discovered, oil companies want to produce them at once to recover their capital, and there is a social cost in postponing production as well.

In the present circumstances the dilemma is confined to the timing of construction of the Mackenzie Valley gas pipeline – whether it should be completed about 1978 or several years later – and to the quantity of oil that should be authorized for export from the Western Provinces during the next few years.

The resources east of Hudson Bay present no problem. Present indications are that Canadian markets could absorb all the oil and gas discovered as soon as it were found. Gas is needed to serve the Atlantic Provinces and Quebec outside Montreal and could be substituted for other fuels in Montreal. Oil is needed east of the Ottawa Valley to replace imports with an assured safe supply. Since oil in commercial quantities has not been discovered yet, it will probably be several years before it will be necessary to decide whether there is an exportable surplus. If oil is not found in the Atlantic offshore fields by say, 1976, it may be necessary to either accelerate exploitation of the Athabasca tar sands to provide oil to replace imports in eastern markets in the 1980s, or to seek a treaty with a South American country to jointly develop its untapped reserves (with the reward to Canadian participation taking the form of assured supplies). In the latter case, the Canadian government might well have to approach a South American government directly, because it is reported that some are reluctant to enter new concession agreements.

The circumstances surrounding the oil reserves in the western provinces and western arctic are rather different. Even in the unlikely event that oil exports from western Canada were shut off after 1974, and the estimated 5 billion barrels of undiscovered reserves in the region were brought into production by 1985, these new reserves combined with the existing proven reserves of 9.7 billion barrels could meet the needs of Canadian markets west of the Ottawa Valley only until about 1989. Thereafter, the excess of demand over supplies available from conventional sources could probably be met to 2020 by production from the tar sands.

In August 1973, the federal government announced its intention to extend the inter-provincial oil pipeline from Toronto to Montreal. This proposal presumably has several objectives: to supply some part of Montreal's needs with western crude; to provide a transportation link for additional supplies in the event of an emergency; and to lift the barrier between prices east and west of the Ottawa Valley so that the lower cost of other supplies would act as a price ceiling. Only limited quantities of oil will be available for Canadian markets east of the Ottawa Valley unless very large reserves are discovered in the arctic or production from the tar sands is increased much more rapidly than is generally considered practicable.

If the G.S.C. 1973 estimates turn out to be tolerably accurate, the only source of large exports west of Hudson Bay will be the heavy oil sands, and then only if the price of oil rises above $6 per barrel (in dollars of 1972 purchasing power).[3] The G.S.C. estimates indicate no big pool of oil in the Canadian arctic at all comparable to the 9 billion barrels at Prudhoe Bay in Alaska. At a delivered cost of $8 per barrel, the largest pools indicated are in the Sverdrup Basin (2.9 billion barrels) and the Beaufort-Mackenzie area (2.6 billion barrels). The total potential reserves estimated to be available from northern Canada west of Hudson Bay at that price amount only to 9.8 billion barrels (and only 6.8 billion at a delivered cost of $6.00). These estimated reserves are small in relation to the 30 billion barrels estimated to be recoverable from the heavy oil sands at a delivered cost of $5 per barrel and the additional 30 billion barrels at a cost of $6 to $8.00 per barrel.

The most urgent decision to be made in evolving an oil policy suited to the new circumstances of world oil scarcity concerns the volume of western Canadian crude oil to be exported during the next few years. As mentioned above, the 9.7 billion barrels of proven reserves together with the 5 billion of potential reserves that may be discovered by 1985 are sufficient to meet Canadian demand west of the Ottawa Valley to about 1989; thereafter, conventional production will have to be supplemented by increasing production from the heavy oil sands. If the NEB follows the rule that enough proven reserves must be maintained to meet domestic consumption for fifteen years ahead, as much as 1.5 billion barrels might be assigned to Montreal and to exports during the next few years. Just how much should be allocated to these other markets must depend upon estimates of how much oil sands production can reasonably be expected during each of the next ten to twenty years. In any event, exports must be sharply reduced.

The industry will presumably press for maximum exports, and their calculations of supplies available may be optimistic. But it is in the interests of consumers to press for maximum security of supplies, and to use conservative estimates of the time at which new supplies will be available from the potential reserves in the west and from tar sands. Because the benefits to Canadians from exports are negligible (see below), it is my view that exports should be confined to the physical surplus of 'assured supplies' to 2010 over domestic consumption. And no potential reserves should be considered as assured supplies until drilling results justify a high degree of confidence in the estimates of quantities available.

The sharpest conflict of interest in prospect for the next few years relates to natural gas, not to oil. The conflict focuses upon the timing of the projected Mackenzie Valley gas pipeline – should it be completed by about 1978, as the Consortium pressing for authorization would prefer, or would Canadians gain greater benefits if it were postponed several years? Some quantitiative information on this question is presented in the next Chapter: but some general observations can be made here.

From a Canadian point of view, it is questionable whether the pipeline should ever be built solely to export the estimated 30 trillion cubic feet (t.c.f.) of gas reserves in the Mackenzie Delta and to trans-ship the estimated 26 t.c.f. of gas in the north slope of Alaska, even if the Delta reserves were surplus to Canadian requirements. But a prior consideration is whether any Canadian reserves should be categorized as exportable surplus before substantial reserves in other basins of the arctic are confirmed. If new reserves are not forthcoming, it would certainly be unprofitable to export existing reserves, because the direct benefits to Candians from exports are negligible, and because exporting now would mean that Canadians would have to turn sooner to higher cost supplies from coal gasification or to oil from the tar sands.

So the issue reduces to the degree of security to be provided to domestic consumers. More specifically, it is the choice between (i) building the pipeline now, exporting perhaps 10 t.c.f. or one-third of the estimated reserves in the Delta over the years following 1978, and depending upon supplies from western Canada to serve Canadian markets, or (ii) postponing the pipeline until Delta gas is needed in Canada. My view of this matter is similar to that in the case of oil. Since the benefits to Canadians from gas exports are negligible, none of the Mackenzie Delta reserves should be allocated to export markets until successful discovery wells indicate supplies sufficient to meet domestic demand for the foreseeable future.

B. FRONTIER ENERGY EXPLOITATION IN CANADIAN INDUSTRIAL STRATEGY

To carry this debate further, let us assume that by 1980 new discoveries will be sufficient to assure supplies to meet domestic needs well into the next century and also to sustain considerable exports. We can thus focus our attention on the priority that should be accorded to investment in pipelines for the export of gas and oil. A related issue is the desirability of maintaining strong incentives for investment in exploration.

In some parts of the federal government's recent review of energy in Canada it is implied that pouring large amounts of labour and capital into arctic oil and gas exploitation for export would be justified by the new employment and income that would be generated.[4] The implication is that such development offers the most profitable opportunities for the billions of dollars of investment that would be needed to sustain the expansion of the Canadian economy at the rate required to maintain a high level of employment during the 1980s. Thus expenditures on pipelines should proceed at as high a rate as possible without overheating the economy because, as of 1973, no other investment flow is assured. But elsewhere in the report it is recognized that investment in pipelines would divert capital from other sectors of the economy and that, if necessary, fiscal measures could assure a high level of employment.

Thus, one polar extreme of industrial policy sees the exploitation of arctic petroleum resources as a fiscal tool for maintaining full employment and generating fast economic growth (but probably not increasing national income per capita). This is consistent with the "Stagnation Thesis" that has cast its shadow over all our fiscal policies since World War II. It is assumed that, left to itself, investment will not be maintained at a level high enough to keep the economy expanding fast enough to keep the growing labour force fully employed. Thus, whenever unemployment appears to be a problem, the Department of Finance faithfully unveils yet another incentive to private investment. The alternative of increased consumption spending is viewed with a mixture of apprehension (that it will overheat the economy) and skepticism (that feasible stimulants will be ineffective).

The opposite extreme to this timorous attitude toward our economic future is the view that, in the absence of large economic gains (or rents) to Canadians, the exploitation of frontier oil and gas *for export* has no place in any sensible Canadian industrial strategy. Instead, exports should be the chance by-products of the search for oil and gas to keep Canada self-sufficient (as long as domestic supplies are no more costly than imports). The reasoning that leads to this latter conclusion may be conveniently summarized as a series of arguments:

1. If there is a danger that investment will fall below the amount required to maintain full employment, the most effective remedies are fiscal measures to increase demand and output across all industries. Special incentives for exploitation of oil and gas would be a poor choice, because the employment created directly and the stimulus to other industries indirectly are heavily concentrated in the years when large pipelines are built. If only one major line is built, there is a danger that an investment boom will be generated and followed by a recession – reminiscent of the 1956-57 investment boom and subsequent recession. Alternatively, if several pipelines were built in succession over a number of years there would be a structural change in the economy, as investment induced by oil and gas exploitation displaced investment in other commodity-producing industries. Other export industries and the import-competing industries would be smaller than they would otherwise be. And when the pipeline construction bulge ended it would be difficult for these industries to recapture their lost market shares. Consequently, resort to the exploitation of frontier oil and gas for export could be justified as a means of maintaining growth only if it were the only means to that end. But it clearly is not. Moreover, maintaining economic growth in step with the increase in the labour force is not expected to be as great a problem in the 1980s as it will be in the remainder of this decade because the growth in the labour force will be lower.

Even if a recession were encountered, it would be difficult to use pipeline projects as a means of mitigating the down-turn. Recessions cannot be forecast accurately enough to mesh the timing of a huge investment project with its occurence: the economy might well be recovering just when pipeline construction was reaching its peak, with the result that it might aggravate an inflationary condition.

2. The effects of pipeline construction on the balance of payments would be one avenue by which investment, employment and production of other industries would be reduced. The capital borrowed abroad to finance the project would bid up the foreign currency value of the Canadian dollar, and this would induce increased imports (especially of supplies and components by the import-competing industries) and lead to a decline in investment in export industries. With a flexible exchange rate, the resulting excess of merchandise imports over exports and deficit in the current account would be smaller than the amount borrowed to finance the pipeline, but the Canadian economy would still bear the burden of transferring labour and capital into pipeline and related activities from other export and from import-competing industries.

Thus, once again, we are led to the conclusion that a huge investment project such as the construction of a pipeline is a very poor device to use as a balance wheel to the economy. The capital borrowed abroad to finance it would probably exceed the import content of the project, and the resulting deficit in the merchandise balance of trade would reduce the output of industries that were not stimulated by pipeline construction. An ideal project for reducing unemployment would be no larger than required to take up the slack in the economy without requiring foreign capital financing that causes a deficit in the current account of the balance of payments.

The conclusion that emerges so far is that the exploitation of frontier oil and gas for export and the concomitant construction of pipelines to export markets must stand on their own as investment projects. They should be undertaken only if the capital and labour they require would be more productive in this activity than in other industries. And unless large resource rents are generated in the process, this is not likely to be the kind of activity in which Canada has an international advantage. Pipelines are exceedingly capital-intensive, and the United States has more capital relative to labour than has Canada.

3. There is a third reason why it would be unwise to single out exploitation of frontier petroleum resources for preferential treatment: the 'external' costs are greater than for most other industries. External costs include the inevitable dangers of environmental damage, the large government expenditures on roads,

communications, and other infrastructure, and the significant probability that ghost towns will be left in the wake of exhaustion of reserves. The ultimate decline in activity in the frontier areas should not be ignored. Arctic oil and gas development will induce migration into the northern territories. But when reservoir development and pipeline construction decline, the larger population will be left with an inadequate economic base unless other types of mining develop faster than there is any reason to assume (and even that would only postpone the inevitable decline). In this respect, a lesson might be drawn from Premier Lougheed's concern for Alberta's economic future.

4. Finally, in the absence of large economic rents accruing to Canadians, it is difficult to see how the exploitation of frontier resources for export would significantly increase national income per capita in Canada. Since most of the equity capital would be owned by the international oil companies or their subsidiaries, the returns to capital invested will not accrue to Canadians. Canadians' income will be enhanced only to the extent that construction workers and others employed in the industry might enjoy wage rates higher than they would receive in other activities.

5. But the most serious reservations about promoting frontier oil and gas as the dynamic force of sustained growth relate to the post-construction period when the exports start to flow. At that time, it is entirely possible that the new export earnings would be a net disadvantage to Canadians. Once the wells are drilled and the pipeline systems constructed, the production of oil and gas generates very little employment. If the Mackenzie Valley gas pipeline were built in this decade and gas exported in the 1980s, total employment in Canadian export industries would decrease in proportion to the displacement of exports of manufactured products by the gas exports. Indeed, this is the probable outcome. A major objective of the United States government is to eliminate the deficit in the American balance of payments while the export of capital in the form of direct equity investments is maintained. This requires that the United States earn a surplus on current account. To achieve that objective, the U.S. has been striving for a current account balance with each of its major trading partners, and there is no prospect of a major change in the American trade position. In view of the prospect of increasing expenditures for foreign oil until at least 1985, it is likely that the United States will experience balance of payment difficulties

for that period of time. Also, the new oil and gas policies announced by President Nixon in 1973 should have dispelled any lingering hopes that Canada might be accorded special treatment.

It is prudent to assume, therefore, that Canada will continue to be under pressure for a decade and more to avoid a current account surplus by keeping its merchandise trade surplus no greater than its deficit in the service items of the current account. In consequence, increased exports of gas would have to be matched by a combination of decreased other exports and increased imports. That outcome would be beneficial to Canada only if there were a prolonged investment boom such that the nation was trying to consume and invest more than it produced. And that condition is precisely the opposite of the circumstances implicitly assumed by the advocates of frontier oil and gas exports as the means to sustaining flagging economic growth.

Even if no constraint were imposed by the United States in its trade relations with Canada, the export of gas would probably displace other exports rather than simply add to the total. This is because a current account surplus cannot be sustained unless it is matched by a capital account deficit—a net capital outflow. The required capital outflow would not occur unless the central bank and the federal government adopted monetary and fiscal measures that they have never implemented before and which seem to be regarded with the greatest apprehension. The realistic forecast must be that increased exports of Canadian gas or oil to the United States would be paid for by merchandise imports rather than by income-earning assets.

To add salt to the wound, it is unlikely that the international oil companies that would receive most of the profits from northern resources exploitation would take home their profits. If they did, this would increase the deficit in the service categories of the current account in the balance of payments and permit a matching increase in the Canadian merchandise trade surplus. But it is more likely that the companies would retain their earnings in Canada, to add to their income-earning assets. It is an elementary lesson of economics that an 'unfavourable' trade balance (a surplus of imports over exports) is desirable because it enables a country to consume and invest more than it produces. But the lesson is valid only if the country is simultaneously able to maintain a chosen high rate of employment and if the profits accruing to residents from the capital they borrow exceeds the interest cost on the funds borrowed.

It may be concluded that the objective in exploiting frontier

TABLE I
Petroleum and Natural Gas
Estimates of Potential Reserves and Supply Prices[1]

A. FRONTIER	Transport tariff $ Bbl.	Potential Oil available at a delivered cost of: $5.00 Bbl. $8.00 $15.00 (Billions of Bbl)			Transport tariff c/Mcf.	Potential Gas Available at a delivered cost per Mcf. of: $1.00, $1.30, $1.75 (Trillions of cu. ft.)		
Northern Canada:								
Beaufort-Mackenzie & N.W.T.	.7-1.1	2.5	3.9	5.4	60-75	14	39	70
Sverdrup Basin	.75-1.35	3.0	5.2	7.6	95-100	–	62	137
Other Northern Canada	.75-1.35	1.6	3.0	4.5	100-105	–	–	15
Total Northern Canada		7.1	12.1	17.5		14	101	222
Eastern Canada:								
Scotian Shelf & Slope	Nil	2.2	3.7	5.3	35-40	12	20	26
East Newfoundland Shelf & Slope	.25	8.1	10.5	12.5	60-65	11	26	43
Baffin Island, Atlantic Rise	.75	5.6	7.9	10.2	.75	–	26	49
Other Eastern Canada	.25-.75	2.3	4.0	6.7	15-65	–	12	25
Total Eastern Canada		18.2	26.1	34.7		23	84	143
West Coast Bowser Basin:	Nil	0.1	0.3	0.4	35	2	4	6
TOTAL, ALL FRONTIER[a]		25.4	38.5	52.6		39	189	371

B. WESTERN PROVINCES	Oil (billions of Bbl.)	Gas (trillions of cu. ft.)
Conventional sources:[b]		
Remaining proven reserves as at Dec. 31, 1972	9.7	51.4
Potential reserves – costs not estimated	4.6	43.7
Alberta oil sands		
Recoverable by open-pit mining at cost per Bbl.		
– less than $5.00	15	
– $5.00 to $6.00	20	
– more than $6.00	30	
Recoverable by in-situ methods (the greater part at cost exceeding $8.00/Bbl.	236	
Alberta heavy oils[c] (the greater part at cost exceeding $8.00/Bbl.)	30	

Prices in dollars of 1972 purchasing power.

Sources:
[a] *An Energy Policy for Canada,* Vol. II, Tables 7 and 9, pp. 96 and 98.
[b] *Ibid,* Vol. II, Table 1, p. 32
[c] *Ibid,* Vol. I, p. 90 and Vol. II, p. 73

reserves of oil and gas should not be either to provide a dynamic force for economic growth or to generate export earnings. Nor can the objective be to capture the economic rents. The last follows because the federal government has already given away whatever rents may be generated (see Chapter 5), and because the government's estimates of costs suggest that the rents are not likely to be substantial in any event.

What, then, should the policy goal be? By a process of elimination we are left with the conclusion that oil and gas policy should be an integral part of energy policy, the main objective of which should be to *secure supplies of energy to Canadians at minimum cost.*

TABLE II
Petroleum and Natural Gas
Projected Canadian Demand

	Oil Billions of Bbl.		Natural Gas Trillions of cu. ft.	
	Total in decade	Cumulated total	Total in decade	Cumulated total
1973-1980 incl.	5.2		15	
1981-1990	11.0	16.2	34	49
1991-2000	17.0	33.2	48	97
2001-2010	24.0	57.2	64	161
2011-2020	32.0	89.2	86	247

Assumed Percentage Annual Rates of Increase

	Oil	Natural Gas
1973-1980	5	10
1981-1990	5	4
1991-2000	3.5	3
2001-2010	3	3
2011-2020	3	3

Source: Calculated from *An Energy Policy for Canada* Vol. II, Table 15, "Standard" Forecast of Canada's Primary Energy Consumption, p.29.

FOOTNOTES

1. The Department of Energy, Mines and Resources, *An Energy Policy for Canada: Phase I.* (Ottawa, 1973).
2. The term "probable" here carries its common connotation, not the special definition given to the word by the National Energy Board.
3. When this chapter went to press, the abrupt increase in the cost of imported crude oil (to $10.50 per barrel in March, 1974) was not foreseen. No one knows what the world price of oil will be during the next decade, but few expect that it will remain at the present high level.
4. The Department of Energy, Mines and Resources, *op. cot.,* Vol. 1, chapter 4.

10.
WHERE DOES CANADA'S NATIONAL INTEREST
LIE?– A QUANTITATIVE APPRAISAL

John Helliwell, Peter H. Pearse, Chris Sanderson, &
Anthony Scott[1]

Should Canadians support the construction of a Mackenzie pipeline to link Arctic gas reserves to southern markets? The answer
obviously depends upon what Canadians would expect to gain
from the project. And the gains depend, in turn, upon all the
circumstances that would attend the development – the cost, the
revenue from sales and Canadians' share of them, how much
Canadian gas would be exported, how much Alaskan gas would
be trans-shipped, when the development would be undertaken,
and a host of other factors that have been discussed in the earlier
papers in this volume. The variety of possibilities for development is almost infinite. But it is important for purposes of
national decision-making to assess, as accurately and systematically as available information allows, the potential gains from
alternative courses of action.

Previous essays in this book have dealt with various consequences of building the pipeline. In this final paper, we attempt
to bring quantitative information on these matters together to
indicate how the development of Arctic gas resources would
affect the welfare of Canadians. This enables us to assess the

claims of the Gas Arctic Consortium, as reported in Chapter 2, that some Mackenzie Delta gas will be needed by 1980 for use in Canada, and that trans-shipment of Alaskan gas and substantial exports of Delta gas are required to make a Mackenzie Valley pipeline economically feasible. Our results indicate that none of these claims is justified.

To make an appraisal we have undertaken additional calculations to show the benefits and costs that Canadians could expect to incur.

The surplus of benefits over costs is referred to, in economists' jargon, as the economic "rent" generated by the resource development – a concept already explored by Paul Bradley in Chapter 3. Essentially, economic rent is measured by the value generated by exploitation of the natural resources in excess of all costs, including the value of any other production displaced by this project. It is thus a measure of the net gains afforded by resource development which would not be generated if the resources did not exist or were left undeveloped.

For an informed decision about a natural resource development, it is not sufficient, of course, to show that any particular proposal would generate rents – or net gains. Modifications, great or small, will alter the benefits and costs; thus various alternatives must be compared to find which is best. In this paper we show how the rents change when some of the major features of development, relating to timing, taxes and royalties, exports and trans-shipment of Alaskan gas are altered.

Several of the preceding papers, especially those of Scott and Pearse, and Bradley, have emphasized the "political economy" aspects of energy exploitation – referring to the interplay of producers, consumers, regions and governments in determining the pattern of development and marketing and the distribution of the gains among these groups. It is important to understand not only the magnitude of the rents that can be expected from alternative courses of action, but also who, among the various interest groups, will receive them. Producers will enjoy rents to the extent that they earn profits in excess of amounts they could expect if they invested their capital in other industries; consumers may receive rents in the form of gas prices lower than they would otherwise pay; and government rents take the form of Crown revenues in excess of the revenues that would be received if the Canadian inputs were put to work in other industries. We have estimated how each of these interest-groups would share in the total rents generated under alternative development programmes.

To estimate the potential rents from pipeline construction and Arctic gas exploitation we establish, in Part A below, a pattern of Canadian gas production, consumption, exports and so on that could reasonably be expected if the Mackenzie project were not undertaken. This provides a base against which we can measure the potential gains from alternative schemes of development. Part B describes the main alternative patterns of development that we have chosen to examine, and how the rents to each interest group are estimated. Our results appear in Part C. We also describe supplementary calculations testing the sensitivity of our findings to some of the critical assumptions underlying our calculations. In Part D we compare our findings with the arguments presented by the Gas Arctic Consortium in Chapter 2.

Potential rents are not, of course, the only relevant considerations in deciding upon Arctic resource development: other important effects cannot be measured in the economic metric. Canadians are concerned about possible effects on native people, disturbances to the northern environment, the adequacy of domestic supplies of energy, the influx of more foreign capital, dislocations to the rest of the economy, and other issues. In Part E we examine how these other incommensurable considerations might be combined with our economic findings in determining the national interest.

A. THE BASIS FOR ASSESSING POTENTIAL GAINS

We cannot measure the gains from development of northern gas resources without first establishing what would happen without it. Thus we have estimated the course of events that would transpire if the opportunity for the Mackenzie project did not exist (or were not pursued) to provide us with a base against which we can measure the potential gains from alternative patterns of Mackenzie development.

Prediction of the course of events that would transpire if Delta reserves did not exist, and of all the circumstances that would attend alternative schemes for a Mackenzie pipeline, involves complex problems of theory and calculation. Estimates must be made of the availability of reserves from different sources; the cost of finding, developing and producing them; pipeline costs and tariffs; governmental revenues from royalties, lease payments, and income taxes both in gas production and in other industries; market demand and prices through future time; and all other variables that affect costs and benefits. All these factors must be estimated far enough into the future to cover the planning life of the projects analyzed – in this case to the year 2021. We have made use of a variety of data from private and governmental sources, and have adopted what we consider to be reasonable assumptions where necessary. The data were then analyzed using an extension of the simulation model used by John Helliwell in Chapter 8. There, the model was used to throw light on the general impact of pipeline construction and operation on the Canadian economy through changes in interest rates, the exchange rate, price levels, and so on. Here, the model is supplemented with additional information and equations to calculate the gains and losses that would accrue to each of the main interest groups in each quarter of every year to a terminal date in the next century; and these future streams of rents are then cumulated and discounted to yield their equivalent present value, in 1973 dollars.

The hypothetical base situation is one in which Canada, in the absence of reserves in the Mackenzie Delta, would continue to depend on natural gas reserves in the southern or "non-frontier" regions of the country (largely in the Western Sedimentary Basin of Alberta and the other Western provinces) until the flows available from these areas are no longer sufficient to meet requirements, at which time supplies would be obtained from other sources. We do not attempt to specify the extent to which supplies might be drawn from new discoveries off the Atlantic coast, in the Arctic islands, from abroad through liquification and tanker shipment, from gasification of coal or from other sources. The possibilities are varied and we have no firm basis for distinguishing among them on the basis of their costs. Instead, we have estimated a trend for the cost of supplies from alternative sources which we assume will establish the market value of gas for comparison with the cost of gas from non-frontier regions.

To establish the pattern of gas production in the non-frontier region, it is necessary to have estimates of existing reserves, the rate at which new reserves are likely to be discovered, and the maximum rates at which gas can be extracted from the reserves. The Canadian Petroleum Association reports proved, marketable reserves in the Western Sedimentary Basin of 53 trillion cubic feet (tcf) in 1972.[2] We postulate the continuing exploration effort will add a further 50 tcf in new reserves—which is close to the Geological Survey of Canada's lower estimate of reserves remaining to be discovered in the region.[3] Appendix B explains our assumptions about the rate at which these reserves will be added, and how we have estimated the maximum available flow in each future year by applying a time profile of extraction rates over the lifetime of reserves according to their vintage. Our calculations assume that the cost of new supplies from the non-frontier region will rise because of sharply increasing discovery costs, until by the time the last reserves are found it has risen more than fourfold. We think that this is probably an overestimate of the future costs of non-frontier resources, bearing in mind that the bulk of the past costs of gas were for development drilling, processing plants, and gathering systems, none of which is likely to increase as dramatically as our estimates indicate. We purposely choose to err in this direction, adding strength to our eventual conclusion that continued reliance on non-frontier sources is preferable to immediate development of Mackenzie Delta gas.

Contractual commitments for exports as of 1973 amount to 14 tcf. We assume for our base calculations that all existing export contracts will be honoured, but that no further foreign sales will take place. By examining the terms of these individual contracts we have been able to estimate the quantity that will be exported in each quarter until the last contract expires in 1995.

Canadian domestic requirements are estimated to grow, as forecast for Ontario by the Advisory Committee on Energy, at 8 per cent per year until 1980, then at four per cent annually until 1990, and at 2.5 per cent thereafter.[4] This trend falls between the upper and lower bounds of demand growth recently forecast by the Department of Energy, Mines and Resources (EMR) and is believed to be consistent with the price forecasts in the EMR report.[5]

These supply, demand, and cost assumptions, which are explained in detail in the Appendices, enable us to construct a reasonable picture of the future pattern of gas production, consumption, exports, prices and costs that would be experienced over the next 50-odd years in the absence of a Mackenzie pipeline (as shown in Appendix A). By confronting this scheme with our alternative plans for developing and marketing Mackenzie Delta gas reserves, we can measure the gains that each would generate relative to the common base.

B. MEASUREMENT OF RENTS

From all the possibilities for northern pipeline and gas developments, we have chosen to concentrate on three specific cases.
While this makes our study manageable, it obviously ignores
many variants that deserve consideration. Our three cases are
designed to cover only some obvious combinations of alternatives
with respect to timing, exports and transhipment of Alaskan gas,
in order to demonstrate how the magnitude and distribution of
rents is affected by altering these components of the scheme. Our
first alternative, Case 1, involves development close to the
expected proposal by Canadian Arctic Gas Study Limited, as
described by Earle Gray in Chapter 2. Case 2 involves a short
postponement of pipeline construction, to begin operating five
years later in 1983. It differs from Case 1 also in that all the
Canadian throughput of gas from the Delta is assumed to be
directed to domestic Canadian markets only. Case 3 involves a
longer postponement, with operations beginning in 1988 – the
year in which we estimate that supplies from the southern non-
frontier producing areas will begin to fall short of Canada's
growing requirements. For this case, we assume that the full
capacity of the pipeline will be used to carry only Canadian gas
from the Delta to domestic markets, Alaskan gas being delivered
by other means.

In all our three cases, the physical characteristics of the pipeline are identical to those described in Chapters 2 and 8, with the
same capacity and built in the same way. The economic impact
of its construction is affected only by its timing. As in Chapter 8,
we assume that the cost of labour and capital required to build
the pipeline reflects the values that these inputs could generate in
other productive activities. Thus the pipeline tariffs generate no
rent: all rents derived from exploitation and delivery of the
natural gas resources accrue to private participants at either end
of the pipeline, or to governments.

For each case, we measure the economic rents generated by
producing 26 tcf of gas from the Mackenzie Delta. This is not to
suggest, of course, that this is the extent of reserves in that region
(the Geological Survey of Canada estimates reserves at 120 tcf).
However, it represents half the capacity of the proposed 48 inch
pipeline over its 33 year planning period, and is the quantity of
Delta reserves on which the development proposal is based.[6] The
later development of additional reserves can therefore be treated
as separate projects, omitted here.

We assume that the city-gate value of gas delivered in Central North American markets will rise from a 1973 value of 60¢/mcf by 11 per cent per year to $1.00/mcf in 1978, then at 6 per cent annually to the year 2000, and at 4 per cent thereafter. These prices are in current dollars, reflecting a steady rate of general inflation of 4 per cent per year (so that the corresponding relative or deflated rates of gas price increase are seven, two and 0 per cent respectively). This assumed trend falls in the middle of the range of gas price recently forecast to the year 2000 by the Department of Energy, Mines and Resources.[7]

The values we use for the tariff on a Mackenzie Valley pipeline, the cost of Delta gas, royalties and all the other variables required for our calculations, as well as the way in which they were estimated, are outlined in detail in Appendix B.

To the extent that gas from the Mackenzie Delta and Prudhoe Bay can be found, developed, produced and delivered to markets at a cost less than the market value or the cost of supplies from alternative sources, economic rents are generated. And the extent to which these rents are manifest in producers' profits, Crown revenues and lower consumer prices, determines the distribution of the rents among shareholders, governments and consumers. Equations 1 to 7 in Appendix B show in detail how we have calculated the rents accruing to each of the seven interest groups who share the total net gains. In the paragraphs immediately below, we described the broad features of the rent calculations for each of the seven groups.

1. *Rents to Canadian Governments, From Non-Frontier Production.* Rents accruing to Canadian provincial and federal governments from non-frontier production are measured according to Equation 1 as the sum of lease payments and royalties payable on non-frontier production, less a crude measure of the amount by which taxes on the corporate profits of producing companies fall short (because of the special tax arrangements for this industry) of the taxes that would be generated by equivalent investment in other Canadian industry.

2. *Producers of Canadian Non-Frontier Gas.* Rents to producers of Canadian non-frontier gas are measured in Equation 2 as the difference between the wellhead value of the gas produced and the sum of operating costs, lease payments, royalties and the cost of capital employed in discovering and establishing reserves.

Canadians have a 23 per cent share in these rents, based on the average Canadian equity interest in the Canadian oil and gas industry.

3. *Rents to Canadian Governments From Delta Gas Production.* Development of the gas reserves in the Mackenzie Delta will generate new revenues to the federal government in the form of royalties, taxes on the corporate profits of gas producers in excess of the revenues that would be generated if the capital were invested in other industries, plus any throughput or export taxes. Equation 3 shows how these rents were calculated.

4. *Producers of Delta Gas.* The rents to producers of Delta gas are calculated through Equation 4. These consist of the wellhead price of Delta gas minus operating costs, royalties, corporate income taxes, exploration and development drilling costs and the cost of capital.

5. *Canadian Gas Consumers.* Rents accrue to Canadian consumers to the extent that gas is made available to Canadian markets at prices lower than those of other supplies. Because supplies from the Delta may alter the pattern of consumption of non-frontier supplies, Equation 5 provides for consumer rents from both sources of supply. Rents from non-frontier sources are measured by the extent to which the regulated price of this gas in domestic markets is held below the price that would otherwise obtain (or obtains in export markets) applied to non-frontier production consumed in Canada. Rents from Delta gas consumed in Canada are calculated in the same manner. Less than one-quarter of this consumption is for residential purposes; so most of these consumer rents actually accrue in industrial or commercial uses.

6. *U.S. Producers and Purchasers of Alaskan Gas.* Mackenzie pipeline construction can add to the rents yielded by the Prudhoe Bay reserves to the extent that they can be delivered through such a pipeline at lower cost than by alternative means. Thus the rent to American producers and consumers from trans-shipping 26 tcf of gas from Prudhoe Bay through the proposed Mackenzie pipeline is the difference between the pipeline tariff that would be levied on this system (plus any throughput tax that might be charged by Canada) and the estimated cost of delivery by alternative means. Equation 6 calculates this rent in terms of the

degree to which the wellhead value of Prudhoe Bay gas is enhanced by delivery through a Mackenzie pipeline.

7. *U. S. Purchasers of Delta Gas.* To the extent that Canadian gas from the Delta is exported and sold in the United States at prices lower than the cost of alternative supplies, rents will accrue to American purchasers. Equation 7 defines this rent as the difference between the value of gas in United States markets and the delivered price that American purchasers would pay for imported Delta gas.

C. THE NET GAINS UNDER ALTERNATIVE DEVELOPMENTS

The major distinguishing characteristics of our three alternative cases have already been mentioned and are summarized in Table 1. Case 1 is essentially the proposal which is expected to be filed by the consortium, as described in Chapter 2, involving construction of the pipeline to begin operations in 1978, the throughput capacity being shared equally by Prudhoe Bay and Delta gas, with half of the latter, as well as all the Alaskan throughput, being delivered to purchasers in the United States. Case 2 is similar, though its start-up date is five years later and the Delta gas is directed entirely to Canadian markets. Only brief transitional extensions of existing non-frontier export contracts are assumed to utilize the temporarily excess southern capacity when the northern increment suddenly becomes available for domestic supply. Case 3 involves an independent Canadian pipeline, carrying only Delta gas delivered to Canadian markets, beginning operations when needed to meet Canadian requirements, in 1988.

In Table 2 we present the main results of our rent calculations, all of which require some explanation. The nine rows of the table display answers from nine separate simulations – Cases 1 to 3 each calculated under three different assumptions about the pricing of gas. The seven columns of the table show how the five types of rents to Canadians, already defined, rise above or fall short of those in the base case; thus they also show how the change in total rents is divided.

The two left-hand columns show the changes in non-frontier rents (accruing to governments and producing firms) attributable to the development and domestic use of Mackenzie Delta gas.

TABLE 1
Major Characteristics of the Cases Analyzed

	Pipeline Construction Starts	Year Gas Deliveries Begin	Alaskan Gas Trans-Shipped	Delta Gas Exported
Case 1	1975	1978	26 tcf	13 tcf
Case 2	1980	1983	26 tcf	None*
Case 3	1985	1988	None	None*

* There are some short extensions of existing contracts for export of southern gas to utilize the excess non-frontier production capacity at the time of the addition of the pipeline's output. See Appendix B.

These changes are generally negative because of our assumption that Delta gas gets priority in the Canadian market. Thus all the costs of deferring non-frontier rents because of early development of the Delta (as in Case 1), or because of the large and indivisible block of Delta gas to be digested by the domestic market (especially in Case 3), show up as negative numbers in the first two columns of Table 1. The numbers are derived by subtracting the base case non-frontier rents from the corresponding values calculated for Cases 1 to 3 using equations (1), (2) and (5). Non-frontier rents are high per unit of production relative to those available from the Mackenzie Delta. For example, the base case present values, with no regulation of gas prices, are $19,141 million for producers' rents and $6062 million for government rents. When making comparisons, note also that the non-frontier rents are based on 103 tcf of gas, while the Mackenzie Delta rents relate only to the first 26 tcf of Delta gas.

Columns 3 and 4 in Table 2 are based on equations (3) and (4) showing the rents to governments and producers from the development of 26 tcf of Mackenzie Delta gas.

The fifth column shows rents to Canadian gas users, which arise only if there is some excess current supply of gas (as there was in times past), if there are existing low-price contracts that are binding on producers, or if there is price regulation. Parts A, B, and C of Table 2 illustrate three broad possibilities. In Part A, we assume that contracts for the sale of frontier and non-frontier gas are regularly revised so that consumers pay as much as they would for an alternative energy source. Hence all the additional rents are filtered back to the producers and governments having a stake in the wellhead price. In Part B, we assume that non-frontier gas is priced as in Part A, but that Mackenzie gas is sold according to the wellhead prices and escalation factors in the existing (1973) contracts for the future sale of Mackenzie Delta gas. These prices were also used in the economic impact simulations reported in Chapter 8, and in an earlier study of costs and benefits.[8] Note that only the purchasers of the Delta gas receive any positive rents. In Part C, we assume that the city-gate price of gas is regulated to be 90 per cent of the estimated market value of gas. The difference amounts to 6¢/mcf in 1973, 10¢/mcf in 1978, and grows thereafter at the same rate as the city-gate value of gas. To avoid discrimination between producing fields, and between suppliers for domestic and foreign markets, the price regulation is accompanied by an equivalent tax on gas exports from the Mackenzie Delta and non-frontier fields alike.

The sixth column of Table 2 shows a measure of the total rents to Canadians. It is equal to the sum of the government rents from frontier and non-frontier production (Columns 1 and 3), rents to Canadian gas users (Column 5), and the Canadian share of producers' rents, based on the equity interest of Canadians in frontier (25%) and non-frontier (23%) production. We recognize that the social value of these various components may differ, especially in the eyes of different readers, so that Columns 1 to 5 may be more helpful than the total in Column 6. For example, if a lower gas price leads to wasteful usage, it is clear that our measure of rents to gas users overstates the net advantage. The rents accruing to producers are by definition in excess of the required return, and thus represent a wasteful distribution of the resource rents. In theory, the efficient way of collecting and distributing the resource rents is through taxes and transfers, although no one in a federal system would suggest that the best methods for doing so are easily agreed upon.

The final column of Table 2 shows the Canadian rents in Column 6 as a percentage of the total rents accruing to all parties, given by the sum of the results of equations 1 through 7. Some explanation of the three foreign components of total rent may ease the job of interpreting Table 2. The foreign share of producers' rents is fairly straight-forward, claiming approximately three-quarters of the rents in Columns 2 and 4. Rents to U.S. purchasers of Mackenzie gas arise only to the extent that Delta gas is exported, and at a price lower than the cost of substitutes. One interesting feature of our calculations is that Chicago-based

TABLE 2

Net Rents Due to Exploitation of Northern Gas

(Present values of all future rents, in millions of 1973 dollars)

	1	2	3	4	5	6	7
	Impact on rents from non-frontier production		Rents from Mackenzie Delta Production				
	To Canadian Governments	To producing firms (23% Canadian-owned)	To Canadian governments	To producing firms (25% Canadian owned)	Rents to Canadian gas users	Total rents to Canadians	Canadian rents as % of total rents
A. No regulation of gas prices; all rents accrue in wellhead prices							
Case 1	-113	-217	1502	1489	—	1711	34%
Case 2	- 90	-135	1514	1570	—	1784	43%
Case 3	- 28	- 86	1452	1547	—	1791	62%
B. Delta gas sold under contract prices; non-frontier gas priced as above							
Case 1	-113	-217	-113	-222	1802	1338	27%
Case 2	- 90	-135	-256	-202	3649	3219	74%
Case 3	- 28	- 73	-245	-189	3502	3276	104%
C. Gas prices regulated to be 10% lower (10 ¢/mcf in 1978) than market value in central Canadian markets; plus matching export tax on Delta and non-frontier gas exported							
Case 1	- 94	-124	1522	893	464	2087	42%
Case 2	50	- 56	1024	1045	796	2118	51%
Case 3	230	- 70	1027	1091	607	2121	74%

consumers in Case 1 obtain a total rent of $720 million even when the well-head price is set high enough that Toronto-based gas users obtain no rents. The reason is that it is substantially cheaper to transport gas from Alberta to Chicago than to Toronto, principally because pipeline costs per mile are much higher north of Lake Superior. If in Case 1 the rising prices of Part A of Table 2 were replaced by contract prices, (as in Part B), U.S. purchasers of Delta gas would get additional rents because those prices are substantially below our estimates of market value. Naturally there are no rents to U.S. gas purchasers in Cases 2 and 3, because no Mackenzie gas is exported. The third U.S. component of total rent is that accruing to purchasers or producers of Prudhoe Bay gas. These calculations depend crucially on the unknown costs of taking Prudhoe Bay gas to southern U.S. markets by another route. If those costs were equal to 50¢/mcf in terms of 1973 prices, then the Mackenzie Valley route would generate, in Case 1, rents whose present value is almost 1600 million 1973 dollars. Although the initial price of using the Mackenzie Valley route is almost as high, because of the tariff-setting procedures noted in Chapter 8, a substantial advantage for the Canadian route appears over the life of the project. These rents can only be assessed accurately after better cost estimates become available for the alternative method of shipment. The Canadian percentages in Column 7 are always substantially higher for Case 3 than for Cases 1 and 2, because only Case 3 does not generate U.S. rents from the trans-shipment of Prudhoe Bay gas.

What can be concluded from the intricate calculations behind Table 2? The basic result is clear – there is no net economic advantage to Canadians in development of Mackenzie Delta gas by 1980. The total Canadian rent figures in Column 6 show that Case 3 is in general at least as advantageous as Case 1. Many readers may be suspicious at the similarity of the results for the three cases, and may wonder how robust our answers would be if our estimates were wrong. One of the key factors is the assumed value of gas, so we report in Appendix C some results based on higher and lower gas values. The high and low price series bracket the entire range of future gas prices recently forecast by the Department of Energy, Mines and Resources (see footnote 7). As might be expected, faster growing gas prices increase the rents from frontier and non-frontier gas. The results in Appendix C show that higher gas prices increase rents more for Case 3 than Case 1, with low gas prices having the opposite effect. The differences due to timing and not exporting are not huge even with the extreme price assumptions. Case 3, deferred development, is about $550 million better than Case 1 with the highest gas prices, and $150 million worse with the lowest prices (based on the figures in Column 6 for total rents to Canadians).

We have also re-run Part A of Table 2 using much higher and lower discount factors in the rent equations. A high discount rate means that long-deferred rents are given much lower weight in the present value calculations. At the other extreme, a zero real discount rate would mean that the same present value would attach to a rent however long deferred. Our central value, used in Table 2, is eight per cent per annum in nominal terms, equal to four per cent after allowing for four per cent general inflation. Our high and low values are 12 per cent and four per cent, equal to eight per cent and 0 per cent in real terms. The results show no change in relative advantages of Cases 1, 2, and 3, although at very low discount rates the advantage of Case 3 over Case 1 becomes quite large, about $1200 million.

Finally, we tested the effect on the ranking of the three cases of assuming less rapidly rising costs of developing the remaining non-frontier reserves. We did this by calculating the rents for the base case, and for Cases 1 and 3, on the assumption that' the constant-dollar capital costs of developing new reserves will double rather than quadruple by the time a further 50 tcf have been discovered. This causes the total present value of base-case rents to expand considerably (by about $2500 million, 90 per cent of which would go to the producers) and there is a moderate further increase, about $50 million, in the present value of the advantage of Case 3 over Case 1. The reverse results would hold on the unlikely assumption that non-frontier costs rose faster than our base estimates. Our assumptions about southern costs make little difference to the rankings because we assume in Cases 1, 2 and 3, as well as in the base case, that all of the non-frontier gas is developed and used as fast as is required, after making room for 1 bcf/d of Delta gas in Case 1, 2 bcf/d in Case 2, and 4 bcf/d in Case 3. If the earlier start on Delta gas in Case 1 also encouraged earlier looping to bring additional Delta gas to the Canadian market, then more deferral of non-frontier rents might be involved, thus increasing the relative advantage of Cases 2 and 3.

D. DOES CANADA NEED AN ARCTIC GAS PIPELINE NOW?

In this section, we relate the results of our research to the main arguments presented by Earle Gray in Chapter 2 entitled "Why Canada Needs the Arctic Gas Pipeline." Chapter 8 dealt with the Chapter 2 assessments of the broader economic impacts of the pipeline. Here we shall deal with the points raised by Gray in the Section of his paper entitled "Risks of Deferral."

The key point made by Gray is that United States export markets and the trans-shipment of United States gas supplies are both essential to the viability of the pipeline. Our results indicate that neither export of Delta gas nor trans-shipment of Prudhoe Bay gas is necessary to make a northern pipeline feasible. Our Case 2 involves no exports of Delta gas, and Case 3 entails neither exports nor trans-shipment of Prudhoe Bay gas. Both Cases 2 and 3 are in most circumstances more economically attractive than Case 1. Gray emphasizes that the costs of digesting the Delta gas in Canadian markets are greater if the entire pipeline is ued only to transport Canadian gas to Canadian markets only. This is quite true, and it is natural that he should emphasize the point. But our results indicate that the costs of digesting an additional 4 bcf/d at the end of the 1980's are less than the costs of building a large pipeline years before it is really required.

A second point made by Gray is that if the pipeline is deferred it will require more exports than the Gas Arctic proposal, which allocates 50 per cent of the pipeline to trans-shipping Prudhoe Bay gas, 25 per cent to Delta gas for export, and 25 per cent to Delta gas for Canadian use. But our Cases 2 and 3 involve no exports of Delta gas, and only temporary extensions of non-frontier export contracts that would otherwise be expiring just as the deferred pipelines are coming on stream. The total amount of additional exports assumed in Cases 2 and 3 is minor relative to the amount that Gas Arctic is planning to export in Case 1. Our calculations for Cases 2 and 3 involve total non-frontier export extensions of 2 tcf and 5 tcf respectively, compared with over 13 tcf of exports envisaged in Case 1. In neither Case 2 nor Case 3 are the export extensions essential to the viability of the deferred Mackenzie pipeline. Without the temporary exports, the present value of Canadian rents in Case 2 drops by $120 million and that in Case 3 drops by $310 million.

A third feature of Gray's analysis is that some Delta gas will be needed for Canadian use by 1980. Our calculations indicate that this is not so. Without assuming abnormally high rates of discovery, or of extraction from proven reserves, we calculate that production from non-frontier sources will be sufficient until 1988, and will continue into the next century. It is true, we expect, that increments to non-frontier reserves will be increasingly expensive. But our analysis suggests that we could afford even greatly more expensive non-frontier gas, if it should be discovered, before turning to Delta gas with its very high transport costs. Gray also mentions that early tapping of Delta gas would provide an essential margin of safety in case demand grows faster than projected. We agree that there must be a safety margin, but we think that there are cheaper ways of providing it. One safety margin is provided by the various possibilities for increasing the flow rate of production from proven reserves. Our calculations are conservative in that we make no use of this flexibility, so it remains as a safety margin, to be purchased if required. A second safety factor is provided by the increasing number of large industrial users equipped to shift from one fuel source to another according to price and availability. Finally, the discovery and development rates we have assumed are not very high by historical standards, and could probably be increased with sufficient effort. If, as we suggest, the Arctic pipelines are deferred, and the relative price of gas continues to rise, exploration interest will surge again in the non-frontier regions. Thus it would be a mistake to use the modest non-frontier discoveries of the last three or four years as a basis for pessimistic projections of future possibilities.

In summary, the Gas Arctic position with respect to timing is that construction this decade is necessary to provide gas needed in Canada by 1980, and to take advantage of Prudhoe Bay trans-shipment possibilities. Our calculations indicate that Arctic gas would not be required until the late 1980's, and that the trans-shipment of Prudhoe Bay gas under the presently envisaged terms would not provide any net cost saving for Canadians, relative to a later alternative. The higher the costs of alternatives to the Mackenzie valley route for Prudhoe Bay gas, the greater the potential rents that could be paid to Canada for trans-shipment, thus possibly making Case 2 economically preferable to Case 3. The third set of results in the table in Appendix C illustrate how a throughput charge on Prudhoe Bay gas would provide one way for Canadians to share in possible transportation rents from the Canadian route. In the absence of specific Canadian sharing in these transportation rents (if they exist in substantial size), our results indicate little to recommend Case 2 over Case 3, and even less to recommend Case 1.

E. IMPLICATIONS FOR CANADIAN POLICY

As we emphasized at the outset of this paper, the best interest of Canadians cannot be decided on the basis of calculations of economic gain alone. The data underlying our estimates are inevitably imperfect, and in any event the opportunities for economic gain must be evaluated in the context of other issues at stake.

Our data are drawn from what we consider to be the most dependable sources, our assumptions reflect our judgement about reasonable inferences that can be drawn, and our procedures are those that we believe appropriate for this problem. But other investigators might have chosen different data, adopted different assumptions, and selected different cases for analysis, and so their results would differ. And, of course, unforeseeable future events might leave the most careful estimates wide of the mark. Our information and judgement lead us to believe that our central estimates in Table 2 offer a reasonable basis for comparing alternative developments, and our supplementary calculations lead us to conclude that at least the relative attractiveness of the alternatives is not highly sensitive to our critical assumptions. Nevertheless, the scope of our study is necessarily very narrow, and as information emerges many more alternatives will deserve study. While we have been forced to specify our three cases in detail in order to obtain our quantitative results, they are really meant to be only illustrative of the magnitude and distribution of rents of certain major policy options. For example, the starting date for the pipeline in Case 3 was determined by our estimate of the date at which non-frontier supplies would fall short of Canadian requirements. But our estimate of non-frontier reserves is probably conservative and they may well prove to be adequate for domestic needs until a later date.

Earlier chapters in this volume have dealt with a variety of other consequences of northern gas exploitation. Issues of ecological damage, social dislocation of native people, foreign influence in the domestic economy and so on, all touch on this question, and ought to affect our determination of the national interest.

Integration of these divergent matters into a decision-making framework is awkward because they are incommensurable, but much can be learned by comparing at least the general direction of policy that would provide the best for each. Specifically, we should ask whether these other concerns weigh in favour of early development, in which case we are forced to balance them subjectively against the economic advantage of postponement, or whether they weigh in favour of postponement, in which case our economic conclusions are reinforced.

Our inferences from the work of other investigators in this volume and elsewhere lead us to conclude that these additional considerations generally weigh in favour of postponement. The impact of any development on the natural environment of the north will inevitably be adverse, in the sense that the ecological integrity of the region will be disturbed. Thus, if environmental damage is considered as an economic cost (as it should be), this consideration implies that the net benefits of development are less than those we have calculated. We infer from Dr. Peterson's cautious report in Chapter 7 that the adverse environmental effects of a Mackenzie gas pipeline will be less severe than is often alleged, providing that the precautions that the Consortium promises are carefully taken. We conclude also that the risk of causing serious environmental damage is lessened the longer the project is postponed.

Two of the main apprehensions of ecologists, as Dr. Peterson explains, are the impact of large numbers of people and the present uncertainties about the extent and form of underground ice. During the last couple of years, we have seen an unprecedented burst of scientific interest in ecological conditions in the Arctic. And, for obvious reasons, there has been a parallel explosion of technological and engineering effort to resolve the unique problems of constructing, operating and protecting pipelines in the northern environment. All this has improved our knowledge immensely, but ecological studies, particularly, are not easily hurried. In the sensitive Arctic environment, the ecological responses to disturbance occur very slowly, and the results of experiments in methods of laying and covering pipe can be observed only over many years. Many of the environmental problems have never been encountered before, and we are only beginning to learn how to cope with them. The longer the project is postponed, the more competent we will be in mitigating adverse effects, and the less will be the risk of unforseen damage. We feel confident, therefore, in concluding that the environmental issue, however serious it is, weighs in favour of later rather than earlier development.

Dr. Jamieson's perceptive analysis, in Chapter 6, of the implications of development for northern Indians and Inuits does not tell us whether the net impact will, on balance, be good or bad. But he does explain how the likelihood of improvement can be increased by mitigating adverse effects and by promoting social adjustment. All these opportunities for enhancing the beneficial impact of resource development on northern natives are likely to grow with time. First, the advance in education, modern skills and adjustment to white society is now very rapid in communities that until recently had little contact with the rest of Canada. Even if only this present pace is maintained, the chances of successful sociological and economic adjustment to an enormous and sudden exposure to industrial activity will substantially improve each year.

Second, while research of an anthropological nature has been carried out among native communities for some time, only recently have we begun systematic studies of social adjustment to change. Pilot projects for studying alternative working and living arrangements for natives have been recently begun, for example, and reliable results will take considerable time. With more time the government would have the opportunity to work out more adequate and effective training programmes to enable natives to take advantage of new employment opportunities, not only in pipeline construction but also in more enduring activities that are likely to follow in the wake of development. From the point of view of the rest of Canada, postponement of pipeline construction and increased governmental expenditures on training and community adjustment may mean more subsidies to northern residents from southern Canadian taxpayers in the short run, while in the long run the cost is likely to be less if the programmes are successful in leaving natives more self-reliant and resilient to development.

Finally, a later project would almost certainly find northern people in a better position to take advantage of the potential contribution of the natural resources to the political and economic development of the northern territories. The territories are rapidly being groomed for provincial status, which will carry increased jurisdiction over natural resources and the Crown revenues from them. Representative government is developing apace. And native groups, such as the Inuit Taparisat are only just becoming organized to press the interests of their members. In short, if federal policy toward Arctic energy development puts a high priority on the welfare of native people, there is much to be said for preparing them for it first – not just by delaying the project but by accelerating adjustment programmes. And if the natural resources are seen as a means of advancing the economic and political maturation and self-sufficiency of Canada's north, there is obviously a need for establishing beforehand the mechanisms that will allow these areas to take full advantage of the potential gains.

The prospect of continued economic growth in Canada will bear on the implications of timing for other important concerns. First, the longer this enormous project is delayed, the larger and more resilient will be the economy that must digest it. Thus the dislocations of large capital inflows, imports, and exports, on prices, the exchange rate, employment and the other effects analyzed in Chapter 8 would inevitably be less in a later period. While the consortium has expressed the hope that half the equity in the pipeline will be Canadian, it has also recognized that achievement of this goal might be frustrated by the current capacity of Canadian financial markets. We would expect some increase in Canadian participation in the debt and equity financing if the project were delayed for a decade or so. Third, as we have shown, a later project would eliminate Canadian dependence on export markets for natural gas to meet the cost of pipeline construction. This would, among other things, reduce the displacement of other Canadian exports through exchange rate effects, as described in Chapters 8 and 9. While we do not fully share the apprehensions of some others about the adequacy of Canadian gas reserves to meet future domestic requirements, it should be noted, also, that postponement of the pipeline would stimulate discovery and development of new supplies in the Western provinces and at the same time permit Arctic reserves to be directed to domestic needs.

One final aspect of timing deserves mention. A conspicuous feature of the public debate about Arctic energy exploitation so far is its concentration on the procedures for public decision making. We have witnessed continuous argument about the role of northern autonomy in general and native rights in particular, the extent and nature of public hearings, and the relative powers of provincial and national governments and agencies. A parallel debate has centered on the federal policy guidelines for northern pipelines, and on land use regulations. It is clear that resolving these conflicts, paying due heed to all the interested parties, will not be a rapid process. How best to proceed will become slowly evident.

All of these non-economic considerations tend to favour later development. Our economic assessment of costs and benefits supports this conclusion by showing that construction of a pipeline in this decade does not provide any economic advantage over the deferred alternatives. We therefore conclude that Canada's national interest lies in postponement of northern pipeline construction until the 1980's.

FOOTNOTES

1. The authors are indebted to Hartley Lewis for helpful comments, and to Robert McRae for programming assistance. The model and data used for this chapter are being frequently revised as part of a continuing research project. Revisions between August and December, 1973, altered the data and model in several respects, but left the basic pattern of results the same.

2. The Canadian Petroleum Association, as quoted in *The Oil and Gas Journal* (March 26, 1973), p. 169.

3. The Department of Energy, Mines and Resources, *An Energy Policy for Canada: Phase 1,* Vol. 1, (Ottawa, 1973), p. 91, reports the lower GSC estimate as 46 tcf. The upper estimate is 87 tcf higher. See also Energy Resources Conservation Board, *Reserves of Crude Oil, Gas Natural Gas Liquids and Sulphur: Province of Alberta,* part 4 (Calgary, December 1972).

4. Advisory Committee on Energy, *Energy in Ontario: the Outlook and Policy Implications,* (Toronto, 1973).

5. The Department of Energy, Mines and Resources, *op.cit.,* Vol. 1, Ch. 3, p. 71.

6. Our Case 3 involves using the whole capacity of the pipeline for Delta production, thus requiring twice the reserves of Cases 1 and 2. We have therefore divided our results for producers' and government rents in Case 3 by 2.0 and altered the consumer rents equation to include rents on a maximum of 26 tcf of Mackenzie Delta gas. Simple division by two for the producers' and government rents is appropriate because the costs of absorbing the temporarily excess producing capacity when the Mackenzie pipeline comes on stream are assumed to fall entirely on non-frontier production.

7. The Department of Energy, Mines and Resources, *op.cit.,* Vol. 1, p. 68.

8. The earlier simulation results are reported in "More on the National Economic Effects of Arctic Energy Developments", published as an appendix to the *Minutes of Proceedings and*

APPENDIX A
**Estimated Flows of Natural Gas Required for Export and
Domestic Consumption, and Available Non-Frontier Supplies,
1973-2021**
(Billion cubic feet per day)

Year	Requirements For Exports (EXGASNF)	Domestic[2] Requirements (DEMAND)	Total Requirements	Requirements Minus Non-Frontier Supplies (DEMAND + EXGASNF -GASMAX)
1973	2.8	4.1	6.9	0
1974	2.8	4.4	7.2	0
1975	2.9	4.8	7.7	0
1976	2.9	5.2	8.1	0
1977	3.0	5.6	8.6	0
1978	3.0	6.0	9.0	0
1979	3.0	6.5	9.5	0
1980	3.0	7.0	10.0	0
1981	3.0	7.3	10.3	0
1982	2.7	7.6	10.3	0
1983	2.6	7.9	10.5	0
1984	2.6	8.2	10.8	0
1985	2.5	8.5	11.0	0
1986	2.3	8.9	11.1	0
1987	1.9	9.2	11.1	0
1988	1.9	9.6	11.5	.4
1989	1.7	10.0	11.7	.7
1990	.6	10.4	11.0	.3
1991	.4	10.7	11.1	1.5
1992	.2	10.9	11.1	2.5
1993	.2	11.2	11.4	3.4
1994	.2	11.5	11.7	4.3
1995		11.8	11.8	5.5
1996		12.1	12.1	6.4
1997		12.4	12.4	7.4
1998		12.7	12.7	8.3
1999		13.0	13.0	9.0
2000		13.3	13.3	9.9
2001		13.6	13.6	11.9
2002		14.0	14.0	12.0
2003		14.3	14.3	12.5
2004		14.7	14.7	13.1
2005		15.0	15.0	14.0
2006		15.4	15.4	14.7
2007		15.8	15.8	
⋮		(increasing by 2.5% per year)		
2021		22.4	22.4	

Evidence of the House of Commons Standing Committee on National Resources and Public Works, Issue No. 22, June 5th, 1973, pp. 41-80. The earlier study was different from Case 1 in Part B of Table 2 in that all of the Mackenzie gas was assumed to be exported. Since the net rents to producers and governments are negligible under the existing contract prices, the results showed that almost all the potential rents would accrue to non-residents.

Footnotes to Appendix A

1. Assuming no new export contracts beyond commitments in 1973. Flows derived from terms of existing contracts. Extensions of some expiring contracts were assumed for our simulations of Cases 2 and 3. In Case 2 exports were maintained at 2.7 bcf/d from 1983 to 1989, and 1.8 bcf/d in 1990 and 1991, returning to the base numbers in 1992. In Case 3, exports were maintained at 2.3 bcf/d from 1987 to 1993, 1.8 bcf/d in 1994, 1.2 bcf/d in 1995 and 1996, dropping to zero thereafter.
2. As forecast by the Advisory Committee on Energy, *Energy in Ontario: The Outlook and Policy Implications,* (Toronto, 1973).

APPENDIX B

This appendix contains four sections:
1. Rent equations
2. Definition of other endogenous variables
3. Definition of exogenous variables
4. Definition of coefficients

1. RENT EQUATIONS[1]

The first term in each equation brings forward the rent cumulated to the end of the previous period. At the end of each simulation run, the cumulated rents are discounted back at the same rate (A1894) to present values in terms of 1973 dollars.

1. Non-frontier gas rents accruing to Canadian governments:
   KRNFG)=(1. A1894)JIL(KRNFG$) (A1906 A1904(1 −
   .67A1884)(PGASNF))(.9125GASPRO)−
   .33A1895(PEXOG)(RESCOST) + (PGAS−
   PGASREG)(.9125EXGASNF)

2. Rents to producers of Canadian non-frontier gas:

$$\text{KRENTNF\$} = (1. + \text{A}1894)(\text{JIL})\text{KRENTNF\$}) +$$
$$.9125(\text{GASPRO})[(\text{PGASNF})(1 - \text{A}1904(1. -$$
$$.67\text{A}1884)) - (\text{A}1909)(\text{PEXOG}/1.48)] - [\text{A}1890$$
$$+ .09125(1.05\text{GASPRO})/\text{JIL}(\text{RESBASE} -$$
$$6.)](\text{RESCOST})(\text{PEXOG})$$

3. Rents to Canadian governments from Delta production:

$$\text{KRENTG\$} = (1. + \text{A}1894)\text{JIL}(\text{KRENTG\$}) + \text{TCGAS} + \text{TROYAL} -$$
$$\text{A}1895[(\text{KGAS})(\text{PEXOG})] - \text{A}1869(\text{A}1868)(\text{LFREE}) +$$
$$.9125(\text{PGAS} - \text{PGASREG})(.889\text{EGASD})(\text{EGASXD}) +$$
$$.9125(\text{PGAS})(\text{EGASFLOW} - .889\text{EGASD})$$

4. Rents to producers of Delta gas

$$\text{KRENTP\$} = (1. + \text{A}1894)\text{JIL}(\text{KRENTP\$}) + \text{YGAS\$} - \text{CGASP\$} -$$
$$\text{TROYAL} - \text{TCGAS} - \text{EXPLORE\$} -$$
$$\text{A}1890[(\text{KGAS})(\text{PEXOG})] + \text{A}1868(\text{LREE}) -$$
$$\text{A}1885(\text{EGASD})$$

5. Rents to Canadian gas consumers:

$$\text{KRENTC1\$} = (1. + \text{A}1894)\text{JIL}(\text{KRENTC1\$}) + .9125(\text{PGAS} -$$
$$\text{PGASREG})(\text{GASPRO} - \text{EXGASNF}) + .9125(\text{PGAS}$$
$$- \text{PGASDEL} - \text{A}1901(\text{PEXOG}/1.48)(\text{EGASD})(1. -$$
$$\text{EGASXD})(.889)$$

6. Rents to U.S. producers and purchasers from trans-shipment of Alaskan gas through the Mackenzie pipeline:

$$\text{KRENTA2\$} = (1. + \text{A}1894)\text{JIL}(\text{KRENTA2\$}) + (- \text{TARIFF} -$$
$$\text{A}1899(\text{PGAS}) + (\text{A}1900 - 11.)(\text{PEXOG}/$$
$$1.48)(\text{EGASFLOW} - .889\text{EGASD})$$

7. Rents to U.S. purchasers of Delta gas:

$$\text{KRENTD2\$} = (1. + \text{A}1894)\text{JIL}(\text{KRENTD2\$}) + .9125[(\text{PGASREG}$$
$$- \text{PGASDEL} - 11.(\text{PEXOG}/$$
$$1.48)](.889\text{EGASD})(\text{EGASXD}) - (\text{A}1868)(1 -$$
$$\text{A}1869)(\text{LFREE})$$

2. DEFINITION OF OTHER ENDOGENOUS VARIABLES USED IN THE RENT EQUATIONS

The equations for variables marked with an asterisk (*) are located in the appendices to Chapter 8.[2]

COSTNF   marginal wellhead cost of non-frontier gas, in ¢/mcf

	=	A1907 (PEXOG)[GASACUM + RESBASE - A1908](A1902) + A1909 (PEXOG/1.48)

CGASP$ operating expenses in Mackenzie Delta gas production.*

GASPRO   gas production from non-frontier sources

    = DEMAND + EXGASNF-(1. -EGASXD)(.89)(EGASD). If GASPRO exceeds maximum production (GASMAX), then GASPRO = GASMAX.

KGAS   The constant 1961 dollar value of the net stock of gas production plant and equipment*

PGASD   The wellhead price of Mackenzie Delta gas, in ¢/mcf. The variable is used to replace the exogenous contract price series EPGAS for all simulations except those reported in Chapter 8 and in Part B of Table 2.

    = [PGASREG - TARIFF - A1901 (PEXOG/1.48)]/1.125

PGASDEL   the delivered price of Delta gas at a southern terminus of the proposed Mackenzie Valley pipeline*

PGASNF   the wellhead price of non-frontier gas, in ¢/mcf

    = PGASREG - [1.16(A1901) + A1899](PEXOG/1.48)

RESBASE   the net stock of proven non-frontier gas reserves in tcf

    = JIL (RESBASE) + DISCOV - 0.9125GASPRO

RESCOST   the cumulated cost of the existing stock of non-frontier reserves in millions of 1961 dollars.[3]

    = JIL (RESCOST) [1.0 - 0.09125 (1.05) GASPRO/JIL (RESBASE - 6.)] + A1907 [RESBASE + GASACUM - A1908](DISCOV)

TARIFF   tariff charge on the Mackenzie Valley pipeline*

TCGAS   corporation tax on Delta gas production*

TROYAL   federal royalties on Delta gas production*

YGAS$ producers' receipts from sales of Mackenzie Delta gas

3. DEFINITION OF EXOGENOUS VARIABLES

Variables marked with an asterisk (*) are described in the Appendices to Chapter 8.

DEMAND requirements for domestic consumption, estimated to grow, as Ontario's Advisory Committee on Energy forecasts for Ontario, at 8% per year until 1980, at 4% annually between 1980 and 1990, and at 2.5% per year thereafter.[4] Average annual requirements in bcf/d are shown in Appendix A.

DISCOV   quarterly additions to the non-frontier stock of proven
gas reserves, in tcf. Discovery rates were chosen separately
for the base case, and for cases 1, 2, and 3. In each case
the objective was to minimize the stock of excess producing
capacity, subject to the constraint that annual totals for
DISCOV should not fall below 2 tcf or above 3.5 tcf. For all
cases DISCOV is set at .5 (equal to 2.0 tcf per year) until the
end of 1978. The base case value is .875, or 3.5 per year
thereafter. For case 1, DISCOV remains at .5 until the end
of 1983, when it rises to .875. For case 2, DISCOV is .875
from 1979 to 1982 inclusive, and .875 thereafter. For case
3, DISCOV is .875 from 1979 to 1987 inclusive, and .875
thereafter. DISCOV becomes 0 in all cases when the total of
DISCOV from 1973 onwards exceeds 50 tcf.

EXGASNF   non-frontier gas exports, in annual average bcf/d.
The amounts to be exported each year under existing contracts were derived by examining individual contracts, and
by extrapolating from the historical relationship between
actual export volumes and maximum volumes specified in
the export authorization.[5] The annual average values, in
bcf/d, appear in Appendix A.

GASMAX   the maximum available flow from non-frontier gas
reserves, in bcf/d. Two important assumptions underlie the
calculation of these maximum flows. First, all reserves discovered in Western Canada before 1970 were assumed to
be exploited at an average initial rate of 1 million cf/d per
8 bcf of initial reserves, declining after 15 years by 15% per
year for eight years and then producing at tiᴜ resulting
constant rate until 95% of the initial reserves are produced.[6]
Post-'70 reserves are assumed to be produced at an initial
rate of 1 million cf/d per 7.3 bcf of initial reserves for 14
years, thereafter declining at 15% per year for eight years
and then producing at the resulting constant rate until,
again, 95% of initial reserves are withdrawn. This higher
rate reflects trends in recent export contracts; almost all of
which have been based on such higher rates.

Second, since our postulated production rates resulted in
substantial excess capacity in the past, it was assumed that
reserves were not brought into production until required.
Thus, for example, reserves discovered in 1973 would not
be required and therefore would not be produced, until
1977. Non-frontier reserves were being brought on stream
in the year they were discovered.[7] Appendix Table A-1
shows GASMAX for the base case, assuming a discovery rate

of 2 tcf per year until the end of 1978, and 3.5 per year until 1989, by which time the remaining reserves have been discovered. The GASMAX patterns for the other cases have lower values in the early years and higher values later, because of their delayed discovery patterns.

LFREE   stock of interest-free loans from U.S. purchasers to producers of Mackenzie Delta gas.* Because interest on these loans is a part of contract negotiations, the *LFREE* series is set equal to zero unless the existing contract prices are used. In that case, the *LFREE* values are as described in Appendix B of Chapter 8.

PGASREG   the regulated price of natural gas in Canadian markets. Assumed to be equal to *PGAS* in most simulations, but set 10% lower than *PGAS* for the simulations reported at the bottom of Table 2.

4. DEFINITION OF COEFFICIENTS

Coefficients marked with an asterisk (*) are defined in the appendices to Chapter 8.

NUMBER	Value	Meaning
A1868	.0225	quarterly proportional interest rate on U.S. bond issues*
A1869	.15	rate of withholding tax on interest payments abroad
A1884	.46	rate of corporation income tax on taxable pipeline profits
A1885	2.886	quarterly depreciation for book purposes*
A1890	.0286	average supply price of capital to the Canadian business sector*
A1894	.0194	quarterly equivalent of annual time preference rate of 8%
A1895	.01586	average proportional quarterly tax return on capital invested in typical Canadian industry*
A1899		throughput tax, as a porportion of the city‑gate value of gas (*PGAS*) in central U.S. and Canadian markets. Where used, the assumed value is .10
A1900	50.	delivery cost of Prudhoe Bay gas to west coast U.S. markets by liquification and tanker shipment in 1973 ¢/mcf

A 1901   24. pipeline tariff from Alberta border to Toronto, in 1973 ¢/mcf. Based on Trans-Canada pipeline rates recently published by the National Energy Board.[8] Our value was derived by subtracting the cost of acquiring and gathering gas to the Alberta border from the Toronto selling price, and adjusting the result to allow for the "front end loading" inherent in the NEB's ratesetting policy (see Chapter 8). This figure is assumed to be constant in real terms (a nominal increase of 4% per year).

A 1902   .26 factor to convert the capital cost of new non-frontier reserves in millions of current dollars to ¢/mcf of eventual production. Derived by assuming a weighted average lag between capital expenditures and receipt of revenues from production of 7 years, and applying a quarterly producers' real pre-tax discount rate of 3.5%

A 1904   .25 proportional royalty charge on the wellhead price of non-frontier gas, approximately equal to the new rates obtaining in Alberta

A 1906   1. land acquisition costs in the non-frontier region in ¢/mcf. Derived from the Canadian Petroleum Association's reported land acquisition costs, pro-rated to indicate the historical cost of acquiring Crown land for gas production (approximately 1¢/mcf)

A 1906   1.96 coefficients in the linear function for capital
A 1908   57.7 costs of discovering, developing, and building processing plants for non-frontier gas. The past and present costs of 38.9 million 1973 dollars per tcf developed (averaged over the first 74 tcf developed) are assumed to rise to 158 million 1973 dollars per tcf by the time an additional 50 tcf have been developed.

A 1909   3. total operating costs in non-frontier areas, in ¢/mcf. Includes cost of operating wells, flow lines and processing plants. Derived from Canadian Petroleum Association figures for 1971, and assumed to remain constant in constant dollars.

Footnotes to Appendix B

1. The constants appearing in the equations in this appendix are:

   1.125   converts pipeline output to input, allowing for gas consumed in transmission

   .889   converts pipeline input to output, allowing for gas consumed in transmission

   .9125   converts ¢/mcf × bcf/d to millions of dollars per quarter

   1.16   inflates the Alberta border-Toronto pipeline tariff (A 1901) to one covering the entire journey from the Alberta producing fields to Toronto

   1.05   multiplied by *GASPRO* to allocate *RESCOST* over the .95 of reserves eventually recovered

   6.0   the unrecoverable portion of ultimate discoveries

   In research subsequent to that reported in this chapter, all reserves and discoveries are measured in terms of marketable pipeline gas. Hence there is no longer any unrecoverable portion of reserves. Copies of the revised model, which is on an annual basis, are available on request.

2. Except where otherwise noted, all historical data concerning natural gas reserves, discoveries, production and exports were obtained from the annual issues of the Canadian Petroleum Association, *Statistical Yearbook* (Calgary, 1960-1971).

3. The initial value of RESCOST was obtained by cumulating the capital expenditures on exploration, development and plant construction prior to 1973, and multiplying the result by the end-'72 value of (REBASE/(GASACUM + REBASE)) to allow for allocation of costs to pre-1973 production. Because most of the CPA data are reported in terms of expenditures on both oil and gas combined, it was necessary to pro-rate the amounts attributable to gas alone. For this purpose, we assumed a fraction of aggregated figures, equal to the fraction of all successful development wells in any year accounted for by successful gas wells. The rationale for this procedure is that drillers usually know whether development wells will strike oil or gas, and so the fraction of total development drilling directed to gas wells reflects producers' intended or desired division of spending between oil and gas.

4. Ontario Advisory Committee on Energy, *Energy in Ontario: The Outlook and Policy Implications,* Vol. 2 (Toronto, 1973), p. 19.

5. The required data are reported in National Energy Board, *1972 Annual Report,* Appendix VI (Ottawa: Information Canada, 1973), pp. 57-66.

6. This time-pattern of production conforms with that suggested in the Department of Energy, Mines and Resources, *An Energy Policy for Canada, Phase 1,* Vol. 2 (Ottawa: Information Canada 1973), p. 80: except that a higher ultimate recovery ratio (95% instead of 87%) was employed to reflect trends in recent contracts.

7. Because our data are drawn from figures of the Canadian Petroleum Association, which defines proven reserves as being limited by the delivery system, it is not necessary to assume a development lag before production can begin.
8. National Energy Board, *Reasons for Decision: in the matter of the application under Part IV of the National Energy Board Act of Trans-Canada Pipelines Limited,* Appendix IV (Ottawa, May, 1973), pp. 1-3.

APPENDIX C
Net Rents Calculated Under Alternative Assumptions

	1	2	3	4	5	6	7
	Impact on rents from non-frontier production		Rents from Mackenzie gas and pipeline		Rents to Canadian gas users	Total rents to Canadians	Canadian rents as % of total rents
	To Canadian Governments	To producing firms (23% Canadian-owned)	To Canadian governments	To producing firms (25% Canadian owned)			

High value of gas. Similar to part A of Table 2, except that the value of gas (PGAS) rises at 7% per year from 1978 to 2000, and thereafter at 6%.

Case 1	- 53	69	2501	2544		3100	42 %
Case 2	- 27	160	2762	2892		3494	49 %
Case 3	- 15	- 28	2915	3100		3669	61 %

Low value of gas. As above, except that PGAS rises at 5% per year from 1978 to 2000, and thereafter at 2%

Case 1	-151	-399	705	647		624	20 %
Case 2	-124	-299	574	573		525	26 %
Case 3	- 35	-121	421	452		472	66 %

Regulated prices plus throughput tax. Part C of Table 2 supplemented by a throughput tax on Prudhoe Bay gas, equal in size (starting at 10¢/mcf in 1978) to the export tax and the price gap in the two-price system.

Case 1	- 94	-124	2674	893	464	3238	65 %
Case 2	50	- 56	2038	1045	796	3133	75 %
Case 3	230	- 70	1027	1091	607	2121	74 %

Increased royalties on Mackenzie Delta gas. Similar to part A of Table 2, except that wellhead royalties are levied at 25%, about equal to the new Alberta scale

Case 1	-113	-217	2040	950		2115	42 %
Case 2	- 90	-135	2014	1069		2156	52 %
Case 3	- 28	- 86	1904	1095		2130	74 %

CONTRIBUTORS

(Those marked with asterisks are members of the Department of Economics at the University of British Columbia.)

Ernst R. Berndt* is an econometrician whose research has included studies of energy demand in North America.

Paul G. Bradley* is a specialist in petroleum economics and the author of *The Economics of Crude Petroleum Production* and several articles on this subject.

Michael Crommelin is a legal specialist in oil and gas policy, and is currently engaged in research on international comparisons of oil and gas management policies at the Faculty of Law of the University of British Columbia.

Earle Gray is a well-known Canadian petroleum journalist, former editor of *Oilweek* and author of *Impact of Oil* and *The Great Canadian Oil Patch*. He is now Director of Public Affairs, Canadian Arctic Gas Study Limited, Toronto.

John F. Helliwell* is an economist who has done extensive research on the impact of monetary and fiscal policies on the Canadian economy. He is the author of *Public Policies and Private Investment* and a large number of other monographs and research articles.

Stuart M. Jamieson* is a specialist in industrial relations, and in the employment problems of native people. Among his extensive publications are *Industrial Relations in Canada*; *Times of Trouble: Labour Unrest and Industrial Conflict in Canada* and *A Survey of Contemporary Indians of Canada*.

A. Milton Moore* is a Canadian authority on industrial organization and public finance, whose writings inclue *How Much Price Competition?* and *Forestry Taxes and Tenures in Canada*.

Peter H. Pearse* is an economist specializing in natural resource problems, and has written extensively on natural resource policy. Among other affiliations, he is a member of the Commission on Environmental Planning of the International Union for Conservation of Nature and Natural Resources.

Everett B. Peterson has an extensive background in biology and law, and has published numerous articles on ecological problems. Until January 1974 he was Project Manager of the Northern Pipeline Study of Environment Canada and is Co-editor of the Report of the Environmental-Social Committee to the Task Force on Northern Oil Development. Dr. Peterson is now President of Western Ecological Services Ltd., Edmonton.

Christopher Sanderson is a research associate, formerly at the Department of Economics, the University of British Columbia.

Anthony Scott* is a prominent authority on natural resource economics, and is a former President of the Canadian Economics Association and member of the International Joint Commission. His writings include *Natural Resources: the Economics of Conservation*.

Andrew R. Thompson is a well-known expert on oil and gas law, and in addition to many contributions to legal periodicals he is co-author of Lewis and Thompson *Canadian Oil and Gas*, and formerly general editor of *Butterworth's Ontario Digest*. He is a Director of the Canadian Petroleum Law Foundation, a member of the International Council of Environmental Law, Chairman of the Canadian Arctic Resources Committee and President of the Arctic International Wildlife Range Society. Dr. Thompson was recently appointed Chairman of the British Columbia Energy Commission.